TILAPIA FARMING

HOBBYIST TO COMMERCIAL AQUACULTURE
EVERYTHING YOU NEED TO KNOW

DAVID H. DUDLEY, PMP, PE

Tilapia Farming
Hobbyist to Commercial Aquaculture, Everything You Need to Know

Copyright © 2021 by David Dudley, PMP, PE

All rights reserved. No part of this book shall be reproduced, stored in a retrieval system, or transmitted by any means—electronic, mechanical, photocopying, recording, or otherwise—without written permission from the publisher. No patent liability is assumed with respect to the use of the information contained herein. Although every precaution has been taken in the preparation of this book, the publisher and author assume no responsibility for errors or omissions. Neither is any liability assumed for damages resulting from the use of the information contained herein.

ISBN: 978-1-7350055-7-7 (paperback)
ISBN: 978-1-7350055-8-4 (ePub)
Also available for Kindle

Published by: Howard Publishing

CONTENTS

PART I	**Tilapia Aquaculture Fundamentals**	**7**
	CHAPTER 1 Tilapia Aquaculture and the BIG Picture	9
	CHAPTER 2 Toxic Food vs. Healthy Tilapia Farmed Fish	25
	CHAPTER 3 Location and Setup Considerations	49
PART II	**Fish**	**53**
	CHAPTER 4 Fish — Everything You Need to Know	55
	CHAPTER 5 Fish Feed	65
PART III	**Components Used in Aquaculture**	**83**
	CHAPTER 6 Equipment & Component Overview	85
	CHAPTER 7 Fish Tanks and Ponds	91
	CHAPTER 8 Plumbing	103
	CHAPTER 9 Liner Material	119
	CHAPTER 10 Making a Water Tight Container	123
	CHAPTER 11 Pumps & Choosing the Right Pump	127
	CHAPTER 12 Filtration (Mechanical, Biofiltration, Natural)	131
	CHAPTER 13 Alternative Energy Options & Operating Off-The-Grid	141
PART IV	**Operation and Maintenance**	**151**
	CHAPTER 14 Starting, Operating, and Troubleshooting Your Aquaculture System	153
	CHAPTER 15 Water Quality	165
	CHAPTER 16 Fish Breeding, Fish Reproduction, and Raising Your Own Crop of Fish	185
PART V	**Tilapia Business Success**	**211**
	CHAPTER 17 Bartering Your Aquaculture Products	213
	CHAPTER 18 Marketing and Selling Your Fish	217
PART VI	**Appendix**	**227**
	Other Books	228
	Country Breeze Farm	247

DEDICATION

This book is dedicated to my son, Nathan, and my daughter, Hannah, both of whom I love and cherish more than words could ever describe. Thanks be to God for the blessings you are to me.

This book is created for every person, young and old, employed and unemployed, educated and those with little schooling, who desire to eat truly healthy food, dream of becoming better off financially, who genuinely care about the environment, and aspire to build a happy, successful, and rewarding life. It doesn't matter whether you live — in the city, suburbs, or country — this book was written with the hope and intention of enabling you to bring your dreams to fruition, directly where you are in life and location.

If you benefit from this book, would you please leave me a positive review on Amazon? It only takes a moment. Positive reviews are a tremendous help to me and are greatly appreciated.

Thank you SO much!

PART I

Tilapia Aquaculture Fundamentals

CHAPTER 1

Tilapia Aquaculture and the BIG Picture

Tilapia Overview

Tilapia is one of the most widely cultured fish in the world. Tilapia, now the second most farmed fish in the world (behind salmon), has played an important role in the growth of aquaculture and will continue to in the future. It is estimated that aquaculture production will grow by 40% by 2030 to satisfy global fish demand.

Several factors have contributed to the rapid global growth of tilapia. Tilapias are easily cultured and highly adaptable to a widerange of environmental conditions. Tilapia feed on a wide variety of dietary sources, including phytoplankton, periphyton, zooplanktons, larval fish, and detritus. Adult tilapia are principally herbivorous but readily adapt to complete commercial diets based on plant and animal protein sources. In the United States, the most commonly farmed tilapia species are, in order, Nile (Oreochromis niloticus), Mozambique (O. mossambicus), blue (O. aureus), and hybrids (Green, 2006).

Past, Present, Future

U.S. sales of organic food and beverages grew from $1 billion in 1990 to $26.7 billion in 2010. The demand for organic food is expected to increase by 16% a year to an estimated 323.56 Billion by 2024 according to Market Research. Consumers are demanding more from organic food products; they are increasingly looking at ethical sourcing, trace- ability, the carbon footprint, sustainability, and corporate social responsibility when making their food buying decisions.

Tilapia farming is an ideal solution since it is a sustainable healthy food production alternative that aims to conserve and reduce the amount of unnecessary resources. It allows virtually anyone to produce healthy food while maintaining a sustainable way of life, at a low cost. Tilapia farming uses no unhealthy chemicals or hormone treatments, which have also proven to be harmful to people and the environment.

One of the most beneficial aspects of tilapia farming is that anyone can purchase and set up an aquaculture system from home. Although tilapia farming may seem quite complicated at first, it is relatively easy to do after one learnshow to do it and gains a little experience.

This method of sustainable food production does not require substantial water usage. Recycled material such asplastic drums, containers, and pipes are often used as equipment for these systems. This strategy reduces water usage, uses recyclable material, and provides a more sustainable way to produce healthy food at its finest.

Seriously disturbing weather and rain patterns happen globally. Massive crop failures, due to drought, are quite com- mon. Even the so-called 'drought resistant' genetically modified crops are not immune to the major droughts that ravage our planet. Conventional agriculture, whether it be organic, GMO or something in between, depends on reservoirs with increasing use demands and rapidly depleting underground water supplies. The vast majority of fertilizer used in agri-culture today is made from natural gas and oil dependent feedstock. All of these ingredients are finite; meaning they are limited and will keep going up in price until they are eventually depleted. The cost is always passed on to the consumer. In many parts of the world, the cost for basic staple foods already exceeds personal incomes. As a result hunger has become a normal daily experience for many people because food grown conventionally is becoming too expensive, and conventional agriculture is vulnerable to unpredictable weather. Food insecurity and hunger is becoming a reality for everyone.

Furthermore, many billion-dollar corporations are buying, or have already purchased food companies, farms (large and small) huge tracts of land, food distribution companies, and water resources. This practice has been, and is, taking place all over the world—in both developing countries and in rich countries such as the United States.

How, then, are we going to eat? While debates rage and conventional agriculture goes on with business as usual, crops being bombarded with droughts, storms, toxic pesticides, and crop-devastating diseases immune to modern controls, hunger is increasing its grip on growing human populations across the globe. Conventional agriculture is proving totally unsustainable. Global food prices are rising annually. That includes the food prices in your local corner market.

The answer has already been invented and is catching on. It is totally sustainable. It does not use soil, toxic pesticides, herbicides, artificial fertilizers, harmful chemicals, genetically modified crops or fish, or antibiotics. It wastes less water than conventional agriculture methods. It does not need expensive, unhealthy genetically modified seeds or plants. It promises to end food insecurity for millions of people. Tilapia farming can help solve the problem. Since the fish grow in tanks water evaporation is greatly reduced, especially since the fish tanks are shaded. Tilapia farming uses approximately 90 percent less that than conventional agriculture.

Tilapia farming also uses around 17 percent of the energy used by conventional farming, since no trucks, tractors, or other machinery is necessary. As a modest user of energy, it is very suitable to using alternative energy sources such as wind power or solar panels. Working with the plants and harvesting is quick and easy, since everything is done at a comfortable ergonomically correct waist-high level. In summary, tilapia farming is an efficient way to grow highly productive, sustainable, and healthy food.

Tilapia Imports and Associated Dangers

Although wild tilapia are native to Africa, the fish has been introduced throughout the world and is now farmed in over 135 countries. Tilapia production globally has steadily increased over the past decade. Global 2018 production was estimated at nearly 6.3 million metric tons (MT). The U.S. imported around 300,000 MT of tilapia in 2018.

China is the world's largest producer of farmed tilapia, supplying approximately 40% of global production. China produces over 1.6 million metric tons annually and provides the majority of the United States' tilapia imports. According to Monterey Bay Aquarium Seafood Watch, over 95 percent of tilapia consumed in the U.S. in 2013 came from overseas, and 73 percent of those imports came from China.

The U.S. Food and Drug Administration (FDA) on the safety of food imports from China noted that in that coun- try "Fish are often raised in ponds where they feed on waste from poultry and livestock" and cited an increased rate of FDA rejection of fish imports from China. In fact, it was reported by EconomyInCrisis.org that Alabama rejects between 50 and 60 percent of imported seafood; but the **FDA inspects less than 1 percent of our imported seafood.**

Tilapia Are Often Fed Animal Feces

One report from the United States Food and Drug Administration (FDA) revealed that it is common for fish farmed inChina to be fed feces from livestock animals. Although this practice drives down production costs, bacteria like Salmonella found in animal waste can contaminate the water and increase the risk of foodborne diseases.

Imported Tilapia Are Often Polluted with Harmful Chemicals

Another article reported that the FDA rejected over 800 shipments of seafood from China from 2007–2012, including 187 shipments of tilapia. Again, the FDA inspects less than 1% of the tilapia imports, so it is safe to presume that a lot of contaminated fish imports are being consumed by Americans.

It cited the fish did not meet safety standards, as they were polluted with potentially harmful chemicals, including "veterinary drug residues and unsafe additives". Monterey Bay Aquarium's Seafood Watch also reported that several chemicals known to cause cancer and other toxic effects were still being used in Chinese tilapia farming despite some of them being banned for over a decade.

Imported tilapia could cause Alzheimer's and cancer. Imported tilapia can carry up to 10 times the amount of carcinogens as other farm raised fish. This is because of the "food" the foreign farmers (primarily in China) typically feed the fish is feces, pesticides, and industrial-grade chemicals. Additionally, the fish may contain high levels of arachidonic acid, which, in excess, has been linked to conditions like Alzheimer's

Imported Tilapia May Contain the Harmful Chemical Dioxin

Researchers have found that dioxin, which is linked to the development and progression of cancer is found within imported tilapia due to the food farmers feed it. However, that doesn't mean that tilapia as a whole all contains contaminants. It all depends on where and how tilapia are raised.

Imported Tilapia are Often Fed Antibiotics and Hormones

Half of the world's seafood is raised on farms, and some of those fish are bound to get sick at some point. So fish farmers, just like animal farmers, are keen on using antibiotics — sometimes in huge quantities. Farmed fish are given antibiotics to fight disease and growth hormones to accelerate growth and production.

Foreign Owned/Controlled Farmland and Food Supply

Another disturbing practice, occurring at an exponential rate over the last decade, is the purchase of real estate by foreign countries and companies. A slew of foreigners—primarily Chinese state corporations and Gulf sheiks— are buying up farmland throughout the world at an accelerated pace to acquire as much precious soil, farmland, and water as possible. This phenomenon is known as "land grabbing," This practice displaces family farms and drives up food costs. Large companies and foreign countries are rapidly obtaining the ability to control food supply and distribution. Other countries regularly farm USA land and then ship the harvest back to their country. The economic outlook of forthcoming higherfood prices in the near future, because of this practice, is alarming.

Data shows that this troubling trend, of foreign governments with trillion-dollar budgets and large foreign corporate companies with million dollar budgets, purchasing enormous amounts of precious limited farmland. Although this has been occurring since the 70's it has been increasing at an exponential rate over the past couple of decades. According to a May 2019 NPR report, "nearly 30 million acres of U.S. farmland are held by foreign investors. That number has doubled in the past two decades.

Foreign entities are also buying up critical farmland and water source acreage in enormous quantities in Mexico, Central America, South America, Caribbean, Asia, and Africa. Again, the pace and amount of land being grabbed is astonishing and truly disturbing. As of 2013 "water grabbing" by corporations amounted to 454 billion cubic meters per year globally. Cooperate and foreign investors from around developed countries are purchasing water rights in some of the most agricultural and environmentally sensitive regions in the world, as well as in arid places where water is already scare for the region's populations. India, United Kingdom, Egypt, China and Israel—accounted for 60 percent of the water acquired under these deals.

Between 2000 and 2012 nearly two-thirds of the land being purchased was in Eastern Africa and Southeast Asia. During this period over 205 million acres of land were been purchased by foreigners and large corporations. About 62 percent of these deals were in Africa. More than one economist has stated that China is buying Africa to feed its rapidly growing population.

These large land grabs push out small farmers and destabilizes the local economy. In Sudan, for instance, the local population is becoming increasingly dependent on food aid and international food subsidies because the land grabbers are pushing out small farmers, and the produce being harvested is shipped to markets in other parts of the world. Evidence also shows that these large land grabs lead to lost natural ecosystems, as a result of farming at such a large commercial scale. Another problem resulting from these land grabs is the large-scale displacement of local peoples without adequate compensation. These displacements often result in resettlement in marginal lands, loss of livelihoods especially in the case of pastoralists, and the erosion of social networks. Lastly, the reduction of available land drives up land prices and is going to make it all the more difficult for the average person to afford real estate.

Some examples for foreign corporate land purchases include the company Cargill purchased 775,000 acres of Brazil's valuable soybean farmland. Nile Trading and Development purchase of 1,482,632 acres of east Africa's rich farmland. BHP Billiton, a large mining company, purchase of 877,000 acres in Indonesia. Ted Turner of AOL and CNN fame, pur- chased of 111,000 acres in Argentina. The South Korean corporation Daewoo purchased of 1.3 million hectares, half of all Madagascar's agricultural land, to produce corn and palm oil. This is just a small fraction of some of the land grab transactions.

Economics: Food Demand Increases

"The two root causes of our environmental crisis— exploding population growth and wasteful consumption of irre- placeable resources. Over-consumption and overpopulation underlie every environmental problem we face today"
—Jacques Cousteau

Economics and how human behavior could be synthesized down to lines on a graph is fascinating. Take the famous graph of the law of supply and demand. It says that if demand rises and supply stays constant, prices will go up. There are more dollars "chasing" the same number of goods.

There are several main drivers of the projected increase in global demand for food in the next forty years: global population growth, increasing standards of living for developing nations, and depletion of resources.

Population Growth

First, demographers' project the worldwide population will grow from the current seven billion to over nine billion by 2050. The dilemma lies in the fact that we are rapidly approaching our planet's ability to support even the life we have now. We are now consuming planetary resources faster than they are being regenerated, including the planet's ability to process waste.

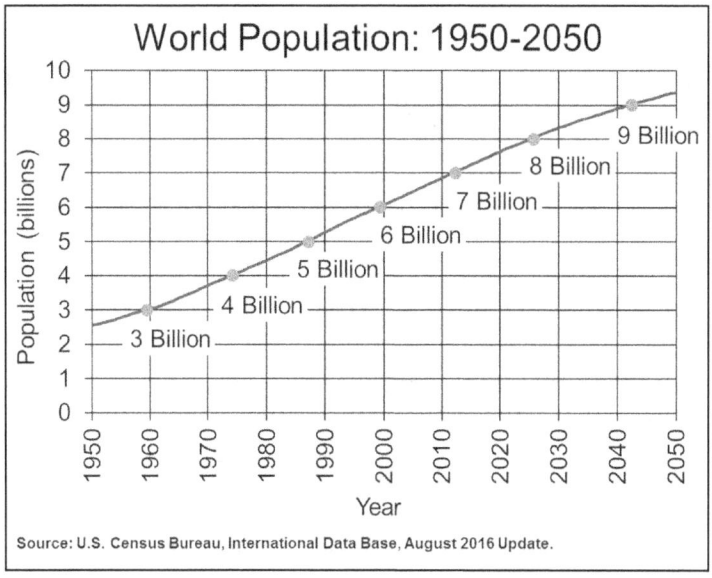

FIGURE 1.

Increasing Standards of Living for Developing Nations

The second big impact on world food demand is globalization. Rising standards of living, especially in China and India where the populations are the highest, will increase demand for a more American lifestyle, especially regarding meat consumption. The technological revolution that is connecting us all is leveling the playing field and opening up a world of possibilities to everyone, no matter what the development status of the country. This opens the eyes of millions across the globe to what a better life looks like, but it brings with it significant change.

Urban areas have more than doubled since 1992. Plastic pollution has increased tenfold since 1980, 300-400 million tons of heavy metals, solvents, toxic sludge and other wastes from industrial facilities are dumped annually into the world's waters, and fertilizers entering coastal ecosystems have produced more than 400 ocean 'dead zones', totaling more than 245,000 km2 (591-595) – a combined area greater than that of the United Kingdom.

This eye opening is driving ever-increasing demand. Consumption is the primary source of strain on the earth's resources. This problem multiplies

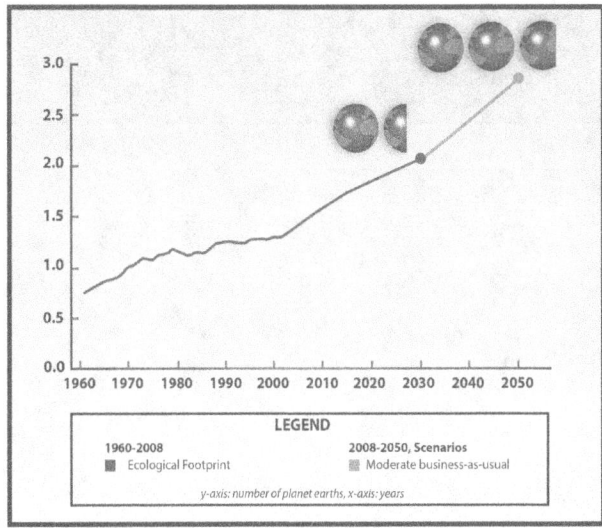

FIGURE 2. Earths needed to sustain population at the current growth rate.

FIGURE 3.

when we try to equally elevate the standards of the developing world. What if other countries were able to increase their standard of living to match that of North America? The Associated Press revealed that each American presently consumes as much as 13 Chinese or 31 Indians. If the Chinese consumed the way we do, we would roughly double world consumption rates. If India and China were to catch up, world consumption rates would triple. If the whole developing world were suddenly to catch up, world rates would increase at least eleven-fold. It is certainly fair that the citizens of the developing world have the right to improve their circumstances, but can the world really afford to have the entire population at the same consumption rate? Imagine if the entire planet suddenly achieved a much higher standard of living.

Economics: Food Supply Decreases

As discussed above there are two reasons global food demand will increase over the next forty years: population growth and increasing standards of living. Based on supply and demand, let's now examine why supply—our ability to grow food at the same rate we have been growing it—is destined to decline.

Our current model of industrial agriculture depends on three factors that are uncertain: inexpensive fossil fuels, political stability, and unlimited water. Following, we will look at how these inputs affects our food supply and show why we should be concerned about our future.

FIGURE 4. Train in India

FIGURE 5. Beach in China

Petroleum Use in Agriculture

As farmers know well, every step in the agricultural process utilizes fossil fuel. From planting (tractors, fertilizers, weed and pest control) to harvesting, factory processing and delivering, all the products and all the steps are dependent upon petroleum. Currently, we use ten calories of energy for every one calorie of food we produce worldwide. This is only sustainable if there is an unlimited supply of cheap, renewable energy. Petroleum, the energy engine of agriculture, is not a renewable resource. While scientists, researchers and activists are debating the exact date of "peak oil," no one disagrees that our supplies are limited. If the supply of something is limited and the demand increases, then the price will rise. Because oil is so intricately woven into every aspect of our current food-production system, increasing oil prices have a direct impact upon the price of food.

All the really easy (and therefore cheap) oil has already been pumped out of the ground. To keep production levelshigh, energy companies are turning to highly disruptive exploration techniques like fracking which use a cocktail of toxic chemicals, and mining oil sand beds, which causes major environmental damage to large geographical areas. Now the energy industry is spending more and more money and resources for every barrel of oil they can recover. When cheap oil collapses, cheap food will no longer be available. The planting, harvesting, storage and transportation of foodrequires enormous quantities of fossil fuels under our current agricultural system.

Other alternative power sources such as wind and wave energy are simply not capable of replacing our current fossil fuel economy—not by a long shot. Solar power is often talked about as a replacement for fossil fuels, but such delusions are mere pipe dreams. For starters, solar panels require rare earth minerals

FIGURE 6.

from China in order to be manufactured. Energy derived from solar panels cannot be easily stored or transported, and there are no electric farm tractors because the energy expenditure of tractors is very high and requires high-density fuels.

Technology is a long way from developing an electric vehicle motor and battery bank that can replace the petroleum-powered tractor engines that run our agricultural system today. Many of today's tractors have engines whichproduce over 200 horsepower.

Water Use in Agriculture

FIGURE 8. Irrigating crop

FIGURE 7. Crop irrigation

Over 70 percent of our Earth's surface is covered by water. Of all that water on earth 97.5 percent is salt water. Only 2.5 percent of the earth's water is fresh water. Of that, less than one percent of the earth's water is actually available forhuman use.

Unfortunately, people are using that one percent at an increasing rate. In a speech on February 5, 2009, United Nations Deputy Secretary-General Asha-Rose Migiro warned that two-thirds of the world's population will face a lackof water in less than twenty years if current trends in population growth, population growth, rural-to-urban migration and consumption continue. Also, interesting to note is that agriculture consumes roughly three-quarters of the world'sfresh water. The proportion is higher in Africa and the United States, with as much as 90 percent of our water use being for agricultural purposes, according to both the United Nations and the USDA.

When the earth had six billion inhabitants, in year 2000, we used nearly 30 percent of the world's accessible, renew- able water supply. Projections for 2025 indicate we will be using 70 percent of the world's accessible, renewable water supply, and like other natural resources, water use will not be evenly distributed. Fresh water is made available to us inthree ways: rain, surface water (lakes, reservoirs, streams, etc.) and near surface groundwater aquifers. All three sources are currently being threatened by overuse and pollution. If the groundwater is not recharged at the same rate that it is being withdrawn, it becomes depleted and eventually disappears.

About one third of the world is completely dependent upon groundwater. As the global population increases, and our need for water rises accordingly, our groundwater supplies will decrease. Already evident in China, groundwater levelsin some areas have dropped at the rate of 5-feet (1.5 meters) per year over the past ten years. According to the Ministry of Water Resources, 90 percent of that fresh water was polluted.

In the United States, 40 percent of our fresh water comes from groundwater supplies. The most famous example is the Ogallala Aquifer, the nation's largest aquifer underlying some 250,000 square miles stretching from Texas to SouthDakota. More than 90 percent of the groundwater pumped from the Ogallala is used for agricultural irrigation. The Ogallala Aquifer is being depleted at a rate much faster than it is being replenished.

Aquifer depletion has drastic consequences that go far beyond the obvious lack of water. The land above a depletedaquifer can turn into a sinkhole and become dangerous and unusable. If an aquifer is close to an ocean, lowering the water level can destabilize the barriers between the aquifer and the salt water. This results in seepage of ocean water into the aquifer, and the remaining water in the aquifer becomes unusable as a fresh water source.

Worse yet, the water we can use is being polluted through the very agriculture that it nurtures. The National WaterQuality Inventory reported that agricultural pollution is the leading source of water quality issues on surveyed rivers and lakes, the second largest source of impairments to wetlands and a major contributor to contamination of surveyed estuaries and groundwater. What is this type of pollution? It is pollution that comes from a wide array of sources insteadof a single "point" like a factory or a sewage treatment plant. Agricultural activities that cause this problem include poorly located or managed animal feeding operations; overgrazing; plowing too often or at the wrong time; and improper, excessive, or poorly timed application of pesticides, irrigation water and fertilizer. Pollutants that result from farming and ranching include sediment, nutrients, pathogens, pesticides, metals, and salts. The consequence is widespread waterpollution and degradation of our lakes, streams and groundwater.

Deforestation

In 2000, nearly 40 percent of the earth's land has already been cleared for agriculture. By 2005, it was closer to 50 percent. The demand for farmland is growing. In our search for more fertile soil, we are turning to the soils of the tropicalrain forests.

According to Rainforest Facts, more than an acre and a half of tropical rain forest is being cleared every second of every day. Experts estimate that the last remaining rainforests could be consumed in less than 40 years.

Unfortunately, by clearing tropical rain forests to create more farmland to feed ourselves, we are destroying valuableecosystems, native plants, and animals at alarming rates, altering natural weather patterns, and destroying a critical air filter our world desperately needs. The rain forests are our greatest source of the

FIGURE 9. Deforestation of native habitat for agriculture. Loss of plants and wildlife habitat. Erosion of topsoil over time.

air we breathe because of their tremendous efficiency in converting carbon dioxide into oxygen. With less rainforests, our condition only worsens.

Fires to Clear Land = Near-Record Loss of Tree Cover

Tree cover loss, mostly in the tropics, totaled 294,000 square kilometres (113,000 square miles) in 2017, just short of a record 297,000 sq kms in 2016, according to Global Forest Watch. Brazil, Democratic Republic of Congo, Indonesia, Madagascar and Malaysia suffered the biggest losses based on satellite data.

Tropical forests were lost at a rate equivalent to 40 football fields per minute in 2017, says World Resource Institute (WRI). Burning of forests to make way for farms from the Amazon to the Congo basin caused a loss of global tree cover amounting to an area almost the size of Italy in 2017.

Vast areas continue to be cleared for soy, beef, palm oil and other globally traded commodities. Much of this clearing is illegal. Unfortunately, until agriculture methods and global consumption habits change the future outlook for forests is not a good one.

The value of agricultural crop production has increased by about 300% since 1970. Raw timber harvest has risen by 45% and approximately 60 billion tons of renewable and nonrenewable resources are now extracted globally every year – having nearly doubled since 1980.

Overfishing the Oceans

Before moving on to tilapia farming as part of the solution, it is important to look at one more critical food supply issue: the overfishing of our oceans. Overfishing occurs when more fish are caught than the population can replace through natural reproduction. In 2015, 33% of marine fish stocks were being harvested at unsustainable levels; 60% were maximally sustainably fished, with just 7% harvested at levels lower than what can be sustainably fished.

Our oceans could arguably be one of the last wild sources of food on our planet, and we are quickly emptying them of fish. Today, 85 percent of the world's fisheries are either fully exploited, over exploited or have collapsed. The global fishing fleet is operating at two and a half times the sustainable level—there are simply too many boats chasing a dwindling number of fish. To date, we have lost at least 2,048 fish species that we know of due to overfishing.

Billions of fish worldwide are killed for food every year. Unless current fishing rates are drastically reduced, scientists predict that every species of wild-caught seafood will collapse by the year 2050. Excessive fish depletion can widely be attributed to the dual culprits of advances in fishing technology and techniques on the one hand, coupled with minimal regulations and restrictions to allow fish to repopulate.

Sadly, some reports state we only eat about 10 percent of all the marine life that is killed in order to feed us. And although other reports are more conservative, saying only about 25 percent of marine life is discarded after being caught, that is still nearly 27 million tons of fish.

Large, computerized ships trawl the deep seas with miles of netting that can obliterate 130 tons of fish in a single sweep. Bottom trawlers cause massive destruction, scraping the sea bottom and destroying miles and miles of coral, sponges, and non-target bottom dwelling fish which are simply discarded as collateral damage. The fact is that the depletion we are seeing is happening because we have become incredibly efficient at harvesting the entire depth of the ocean. Even species once left mostly untouched are now in danger of depletion. We now have the technology and fishing fleet capacity to catch four times the current supply of fish.

In fact, fishing is one of the world's most wasteful and destructive industries. Every year, more than seven million tons of so-called "by-catch", (perhaps more accurately described as "by-kill") is inadvertently caught and unabashedly destroyed; including over sea animals such as non-target fish species, sea turtles, dolphins, whales, sharks, albatrosses, and other sea birds. Every year 7.3 million tons of marine life is caught unintentionally by the fishing industry just to be callously thrown back dead, considered an acceptable loss in the industrial pursuit for profit.

Endangered species are also vulnerable to industry nets and other gear. Just as the dolphins still suffocate and die in the tuna industry nets, sea turtles are killed by the millions in the nets of the shrimping industry. In fact, for every pound of shrimp netted in the Gulf of Mexico, four pounds of "by-kill" is wasted and thrown back overboard dead.

Not only are we depleting the oceans of fish at an alarming rate for human consumption, but half the world's fish catch is fed to livestock! In fact, more fish are consumed by U.S. livestock than by the entire human population of all the countries of Western Europe combined.

To solve this crisis, various governmental agencies have attempted to set annual quotas by species to prevent a complete collapse. For example, the International Commission for the Conservation of Atlantic Tuna (ICCAT) sets the limits on bluefin tuna. However, these limits appear to be based more on politics than science. According to many biologist and other scientist these limits are set far above recommended recovery levels. Furthermore, these limits are largely ignored by commercial fishing companies. The WWF estimates that at least 20 percent of the fish caught worldwide, and as much as 50 percent in some areas, are caught illegally.

Therefore, we need to either stop eating so much fish or turn to aquaculture and farm more of our fish on land so the ocean species can recover. Since the health benefits of fish are so compelling, it is unlikely we are going to eat less fish anytime soon. Fish provide an excellent source of omega-3 fatty acids, vitamins and minerals that benefit a person's overall health. The American Heart Association recommends at least two servings of fish per week to help prevent heart disease, lower blood pressure and reduce the risk of heart attacks and strokes. Compared with other sources of animal-based proteins, all of which are full of saturated fats, fish is the healthy alternative.

Fish are also vastly more efficient sources of protein than other forms of animal protein. Currently 37 percent of the world grain harvest is being used to produce animal protein, and in the United States, that number reaches nearly 70 percent. Unfortunately, most of that gain is going to feed cattle, the most inefficient source of animal protein.

The global demand for fish has increased dramatically. In a recent report on "The State of the World's Fisheries and Aquaculture" the FAO revealed that global average consumption of fish has hit a record of over 41 pounds (18 kg) per person per year.

Unfortunately, some types of aquaculture do not relieve any strain on the supply of ocean fish, since the feed used in aquaculture operations uses ocean-harvested fish to create the fish meal which is its protein base. Furthermore, some large-scale ocean aquaculture operations generate more ammonia and feces waste than many human population centers.

Finally, while you would think that the ability to grow fish in ponds and tanks would be great for local fish production, in the United States only 10 percent of the farmed fish we eat is produced domestically. As of 2020, China produces about 70 percent of the farmed fish in the world, harvested at thousands of giant factory-style farms that extend along the entire eastern seaboard of the country. The transport of this product from halfway around the world results in more unnecessary atmospheric pollution and consumption of limited petroleum resources.

Extinction of Plants and Animals

Our planet now faces a global extinction crisis never witnessed by humankind. Human destruction of the living world is causing a "frightening" number of extinctions, according to scientists who have completed the first global analysis of the issue. Up to one million plant and animal species face extinction, many within decades, because of human activities. In fact, 99 percent of currently threatened species are at risk from human activities.

About 75% of land and 66% of ocean areas have been "significantly altered" by people, driven in large part by the production of food, according to a recently released Intergovernmental Science-Policy Platform on Biodiversity and Ecosystem Services (IPBES) report. Crop and livestock operations currently co-opt more than 33% of Earth's land surface and 75% of its freshwater resources. Agricultural threats to ecosystems will only increase as the world's population continues to grow, according to the IPBES analysis.

Extinction of Wild Plants

Plants are considered a critical resource because of the many ways they support life on Earth. They release oxygen into the atmosphere, absorb carbon dioxide, provide habitat and food for wildlife and humans, and regulate the water cycle. Because of the many ways plants help the environment, their importance should not be forgotten.

Scientist have identified 571 species of plants that have been wiped out since 1750, but with knowledge of many plant species still limited the true number is likely to be much higher. The plant extinction rate is 500 times greater now than before the industrial revolution.

The world's seed-bearing plants have been disappearing at a rate of nearly three species a year since 1900. The number of plants that have disappeared from the wild is more than twice the number of extinct birds, mammals and amphibians combined.

Extinction of Animals

In the past 500 years, we know of approximately 1,000 species that have gone extinct, from the woodland bison of West Virginia and Arizona's Merriam's elk to the Rocky Mountain grasshopper, passenger pigeon to Puerto Rico's Culebra parrot, and many more. The doesn't account for thousands of species that disappeared before scientists had a chance to describe them.

The average abundance of native species in most major land-based habitats has fallen by at least 20%, mostly since 1900. More than 40% of amphibian species, almost 33% of reef-forming corals and more than a third of all marine mammals are threatened. The picture is less clear for insect species, but available evidence supports a tentative estimate of 10% being threatened.

Causes of Extinction

The main cause of the extinctions is the destruction of natural habitats by human activities. Cutting down forests, converting land into fields for farming, overfishing, pollution, and expansion of population centers all put a strain on native plants and wildlife. Below are elaborations of some of the significant causes of extinction.

Agriculture Impact on Extinction

Industrial agriculture disrupts habitats and biodiversity. Habitat loss and unintended chemical runoff are the destructive products of traditional farming practices. Native plants and animals are eradicated on a massive scale.

One example you have likely heard of is the threatened species milkweed, which the vulnerable monarch butterfly relies on for reproduction. Milkweed is killed with agricultural herbicides or mowed away from hedgerows and roadsides. Unfortunately, there are way too many more examples to list.

Beyond the colossal loss of wildlife and native plants, agricultural diversity is also being lost. One estimate suggests that 75 percent of agricultural crops have been lost since 1900. This lack of diversity in food crops is risky for human food security. Agricultural diversity involves not only the production and processing of nutritious food, but also access by individuals to the full range of nutrients needed to maintain an active and healthy life. A greater diversity of genetic agricultural plants also helps to ensure a secure food supply at more stable prices. Greater agricultural diversity also helps prevent massive losses of crops to pests, diseases, and changes in climate conditions.

Habitat Loss

Deforestation and urbanization combine to create two additional reasons why plants and animals become extinct. Deforestation is leveling forests to harvest the wood or create space for building or agriculture, while urbanization is the turning of once-rural areas into cities. As the human population grows, more and more land is cleared and urbanized for living space. This shrinks habitat for animals and plants. Each year, 36 million acres of natural forest is leveled, according to the World Wildlife Fund. The forest provides habitat for 80 percent of the world's species.

Exotic Species Introduction

When animals and plants that are not native to a region are introduced to the ecosystem, they can cause serious damage to the local plants and animals, and contribute to their extinction. Native species must compete with the exotic species for basic needs such as food and water. If the exotic species is more aggressive than the native species, the native species then runs the risk of extinction.

The introduction of the Nile perch into the Lake Victoria ecosystem in Africa represents a prime example of this, according to "Causes and Consequences of Species Extinctions," a paper published by Princeton University Press. The Nile perch was introduced to the area in the 1950s and by the 1980s, a population boom of these fish contributed to the extinction of between 200 and 400 native fish species.

Overexploitation

Overexploitation, also called overharvesting, is the excessive harvesting of an animal or plant species, making it harder for the species to renew its numbers. The Princeton University Press paper points to the Steller's sea cow, which was discovered in 1741, overexploited, and then became extinct in 1768. Save the Frogs, a frog conservation group, notes that several frog species feel the effects of overharvesting for food, pet and scientific purposes. Fish also fall prey to overexploitation. According to Greenpeace, more than 70 percent of native habitats worldwide are either "fully exploited, over exploited, or significantly depleted."

Tilapia Farming Can Be Part of the Solution

The world is getting increasingly more humanized, and it is happening with colossal negative consequences to native plants, wildlife numbers, wildlife biodiversity, agricultural diversity, and wildlife habitat. Native species are being forced into living in smaller patches of habitat given the destruction of natural habitat.; thereby threatening native plants, insects, and animals; causing many to become endangered or extinct.

Tilapia farming greatly helps resolve the problem. Tilapia farming helps save our planet from overfishing. Tilapia farming uses 90% less water than traditional agricultural practices. Tilapia farming also coincides well with areas that have already been developed, such as in cities, towns, and community neighborhoods; thus, allowing more land to be preserved for, or converted to, natural habitat for native plants and wild animals, as well as our recreational enjoyment (e.g., multiuse trails, sportsman activities, camping, etc.).

Differing Views on the Climate Change and Global Warming Debate

There are six common views regarding 'Global Warming'. Although now the most current politically correct catch phrase, is 'Climate Change'. Regardless of what the most outspoken pundits call it, they are referring to the same thing.

Point of View #1: Many believe that global warming started long before the "Industrial Revolution" and the invention of the internal combustion engine. They argue that global warming began long ago, as the earth started warming its way out of the so called 'Ice Age'— a time when much of North America, Europe, and Asia is said to lay buried beneath great sheets of glacial ice- and continues today as part of a natural process.

Point of View #2: Same view as above, adding that the industrial revolution and human activities are now accelerating global warming and climate change.

Point of View #3: The "Industrial Revolution", CO_2 producing human activities, and destruction of our natural resources is the cause of global warming and the world will continue to warm at a devastating exponential rate.

Point of View #4: Global warming is all hype. Planet earth has always experienced variations in climate; and a few decades (or centuries) of warming (or cooling) is just a natural process.

Point of View #5: There is no global warming, it is just bad science and media hype.

Point of View #6: There is no global warming, but the new world order globalist, mainstream media, and government leaders are propagandizing it to push their agenda.

Our Responsibility

Regardless of which view you and I have in regards to the climate change issue, we all share one thing in common. Each one of us has an important moral and ethical responsibility to take good care of this planet during our lifetime. We need to be concerned about the planet, do our part to help out, and reduce our negative impact upon wildlife, oceans, forests, grasslands, and the environment as a whole. We should make every effort to minimize pollution, destruction of natural habitat, consumption of resources, and waste. We should always be striving to do positive things that directly benefit our planet, and the natural environment, regardless of the global warming issue.

A good rule of thumb is to always leave a place (and a person) better off than when you arrived. This perspective, when applied, is a win-win for all. Build others up with an encouraging word, a helping hand, good advice, constructive instruction, and/or an empathetic ear and you will make a positive difference in their lives. This positive impact in the life of another in turn spreads into families and communities like the ripples from a rock thrown into a pond.

Being a good steward of this planet — constantly trying to reduce consumption of natural resources, minimizing our pollution, not over consuming or wasting resources, and making a consistent effort to help the environment — likewise benefits earth's inhabitants now and in the future. Such should not be considered as a good deed, but instead viewed as simply honoring our responsibility.

It has been said that a winner is someone who picks up a piece of trash that is not his or hers. As an individual, we cannot dictate government policy, force companies to operate differently, control others, or convince everyone of our climate change viewpoint, but we can make a big difference in regards to our individual impact upon this planet — for better or worse. Our lifestyle will result in either negative consequences or positive benefits for the wildlife, natural habitat, preservation of species, and quality of life for future generations of this planet. The responsibility is ours — an individual choice that truly makes a difference.

The Aquaculture Solution

Food prices have risen, natural habitats destroyed at outrageous rates; and the supply and demand structures have changed. So, is there hope for the future? While aquaculture cannot address the dual demand pressures of population growth and increasing global standards of living, it does offer solutions to many problems.

Pollution and Traditional Agriculture vs. Tilapia Farming

Tilapia farming is a food-growing system that has an inconsequential impact on our planet compared to traditional agricultural practices, especially if the pump is powered through renewable energy sources. Except for purely wild food-growing systems, such as the ocean, and most permaculture techniques, no other food system can likely make that claim. Furthermore, Tilapia farming can be done where development has already occurred, such as in urban areas, the suburbs, and the residential neighborhoods, preserving wildlife habitat and our natural ecosystems.

On the other hand, as previously addressed, traditional agriculture is the single largest contributor of CO_2 emissions, while simultaneously consuming and requiring more natural habitats for growing crops and raising cattle. The main pollutant sources are CO_2 emissions from petroleum operated equipment used in farm production and food transportation, methane from cattle production, and nitrous oxide from over-fertilizing. Tilapia farming requires none of these inputs. These pollutants are non-existent or negligible in tilapia farming. The need for petroleum is next to nothing (maybe a backup generator on rare occasions). Furthermore, fish do not produce methane as cattle do, and there is no chance of over-fertilizing an aquaculture system.

Perhaps, most importantly, tilapia farming can be started anywhere. This alleviates the need to clear jungles and forests and allows the focus to be with urban areas. Cities and suburbs can become the farms of the future. Aquaculture can produce 50,000 pounds of tilapia per year in a single acre of space. By contrast, one grass-fed cow requires eight acres of grassland. Another way of looking at it is that over the course of a year, aquaculture will generate about 35,000 pounds of edible flesh per acre, while the grass-fed beef will generate about 75 pounds in the same space. It is not an unrealistic notion to think at least some portion of our food can be produced in our urban centers. In fact, it is not unusual for city dwellers to grow a meaningful portion of the food they eat. Hong Kong and Singapore already both produce more than 20 percent of their meat and vegetables within the city limits. With aquaculture, pro- duction yields are much greater, and done correctly have the potential of allowing a city to basically meet all of its animal protein needs. With the projected dramatic decrease of the ocean's fish supply, we must turn to other options to continue enjoying the healthful benefits of eating fish.

Feed Conversion Ratios

Animals must be fed. Raising animals and growing the feed to sustain them takes land and tremendous amounts of energy resources. Obviously, minimizing these requirements is best for our planet as a whole. With such in mind, it is encouraging to see how raising fish has the lowest negative footprint over that of other animal products. Following is alist of the average pounds of feed needed to produce 1 lb. of product for various animal groups:

- Fish = 1.7 lbs
- Chicken = 2.4 lbs
- Turkey = 5.2 lbs
- Pork = 4.9 lbs
- Lamb = 8.0 lbs
- Beef = 9.0 lbs

Summary of Benefits

Tilapia farming is a high efficiency food production method for growing protein dense healthy food year-round, in any climate. Following is an itemized list of some of the benefits derived from tilapia farming:

- Healthy food (100 percent organic food if done correctly).
- Saves money.
- Helps the environment.
- Can be a source of income.
- Harvest can be bartered for other products and/or services.
- A tremendous amount of food can be brown in a relatively small area.
- Uses extraordinarily small amounts of energy.
- Can be easily operated off the "grid" via solar- or wind-power. This is exceptionally important should the utility electrical system become unreliable.
- Enables a person or family to grow their own food year-round for consumption, bartering, or sell to others. This could become especially important should the there be a disruption in commerce.

CHAPTER 2

Toxic Food vs. Healthy Tilapia Farmed Fish

Crop Considerations

Organic food has become a fairly common preference for most people. Despite the growing popularity of organic products there is still a large segment of society, including those who purchase organics, who cannot explain why organic food may be better than non-organic food and do not understand many of the other dangers added to our food: antibiotics, growth hormones, preservatives, arsenic, mercury, and the other 30+ heavy metals commonly found in our foods.

U.S. sales of organic food and beverages grew from $1 billion in 1990 to $47.9 billion in 2019, according to the Organic Trade Association, an industry group. In addition to attracting newcomers, such as Sprouts (a nationwide grocery store chain), the growth in the organic food segment has prompted some traditional grocery stores to rethink their product lineups and are increasingly offering more organic choices. SPINS, a reporting firm which tracks the natural products industry, stated that consumers are becoming more educated regarding the dangers of pesticide use and genetically modified crops. In other words, the demand for organic crops is increasing, and the prices for such continue to rise.

Studies link exposure to many toxins found in our food products to cancer, birth defects, stillbirth, infertility, and damage to the brain and nervous system (including Parkinson's disease). Herbicides, pesticides, and fungicides are designed to kill, and because their mode of action is not specific to one species, they often kill or harm organisms other than pests, fungus, and/or weeds. The application of these toxic chemicals makes its way into our food chain. Even exposure to low doses can cause a range of neurological health effects such as memory loss, loss of coordination, reduced speed of response to stimuli, reduced visual ability, altered or uncontrollable mood and general behavior, and reduced motor skills. These symptoms are often very subtle and may not be recognized by the medical community as a clinical effect. Other possible health effects include asthma, allergies, hypersensitivity, and hormone disruption.

What Are GMOs and Why Are GMOs Bad?

According to the Food and Drug Administration, genetically modified organisms (GMO)—such as plants or animals—have been genetically engineered to create new characteristics. In other words, a specific gene is added to an organism to produce a new trait.

Most of these new unnatural gene alterations allow the plant to be sprayed with pesticides, herbicides, and/or fungicides without being killed. Meanwhile, everything else around the plant dies. Unfortunately, the plant absorbs the toxic chemicals and associated heavy metals, which are then passed on to you and your family. Furthermore, these plants are often used to feed livestock and fish; whereby the toxic substances work their way up the food chain in greater concentrations.

From 2004 through 2017 I lived in the California valley—also known as the Salad Bowl of America since most of the nation's produce is grown here—where it is common to to see a food crop being sprayed with various toxic chemical applications three or four times a season. In addition, with a mild climate farmer are able to grow up to three crops a season, depending on the types of crops being grown. As a result, the same farmland often receives a chemical application many times during the year. These chemicals get more and more concentrated in the soil over time. So, not only are these poisons and associated heavy metals entering the plant through direct application, but they are absorbed by the plant roots from the soil and water.

When we first relocated to California, we were absolutely thrilled to discover such an abundance of fresh fruits and vegetables. Just about every variety of vegetable and fruit can be found at local stores throughout the year. However, it did not take long for our bubble of joy to pop. On just about every drive out of town we saw crops being sprayed or chemicals being added to the water used for irrigation. The amount of chemicals we saw being regularly applied to crops within our area of California alone was truly astonishing. This played a very influential role as to why I took a much bigger interest in growing a lot of my own food.

The California counties known for growing agricultural crops average over 250 tons of pesticides, herbicides, and fungicides applied to crops annually. That is not a typo—over 250 tons of chemicals are applied annually on farms in each of California's agriculture counties!

Yet Florida applies more than eight times the amount of pesticide and herbicides on tomatoes as does California. In order to get a successful crop of tomatoes, the official Florida handbook for tomato growers lists 110 different fungicides, pesticides and herbicides which can be applied to a tomato field over the course of the growing season, and many of those are what the Pesticide Action Network calls 'bad actors' —the worst of the worst in the agricultural chemical arsenal.

The most recent developments in GMO technology actually programs the plant itself to produce toxic substances that repel or kill insects. Genetically modified organisms are found in a huge variety of food products from baby food to fruit juice, and many believe they are dangerous to your health. Washing produce does minimal good, as the plants actually absorb these chemicals into their cells.

FIGURE 10. Crop duster spraying pesticide on food crop.

GMOs Harmful Health Impacts (Bottom Line)

Studies on GMO are almost infinite. Many studies are performed by companies and industries that profit from GMOs, whereas others are conducted by those that are opposed to GMOs. Therefore, getting a true unbiased and untainted representation of the impacts GMOs have on human health can be a challenge. However, if one will honestly examine the evidence produced from 'true' scientific studies, which are not influenced by agenda driven money backers and tainted with biased data collection or have subjective conclusions, it is obvious that GMOs can result in negative health impacts.

Regardless, basic common sense should be enough to discourage us from the consumption of GMO foods. It does not take a brain scientist, a university study, or a rocket engineer telling us that eating foods sprayed with harmful chemicals is not healthy. Why would any reasonable person want to eat food that contains toxic pesticides, herbicides, and/or fungicides? An excellent resource for accurate scientific evidence on the negative impact GMOs have on human health is the Organic Consumers Association—a non-profit 501(c)3 public interest organization campaigning for health, justice, and sustainability (http://www.organicconsumers.org).

The website "Natural News" (http://www.naturalnews.com) also has many articles based upon unbiased science which provide the real truth about GMO foods. Most people who become educated about the harmful effects of GMO foods find themselves becoming non-GMO advocates.

FIGURE 11. Pesticides being sprayed on an orchard.

GMOs Harmful Environmental Impact

In addition to the potential health risks, there is solid evidence that GMOs are bad for the environment. The Union of Concerned Scientists (UCSUSA) mentions six ways GMOs might have a negative environmental impact.

1. GMO crops could become weeds.

2. New genes could move to wild plants, causing those plants to become weeds.

3. Crops that produce viruses may lead to new, stronger viruses.

4. GMO plants created to release toxins threaten wildlife.

5. GMO crops could disturb the eco-system in an unpredictable way.

6. GMO crops threaten crop diversity.

The UCSUSA also states there is no consistent monitoring program in the United States, and there may be further negative impacts yet to be detected.

Harmful GMOs Everywhere

Eighty percent of the food supply in North America contains GMOs. Currently, there are no national labeling requirements for GMO identification. While you may associate GMOs with only corn or soy products, they are found in many others, especially processed foods, which includes, but is not limited to those listed below.

- Vegetable oils
- Lecithin
- Flavor enhancers
- Cereals
- Sugar
- Dairy products
- Beets
- Papaya
- Squash
- Coffee
- Artificial sweeteners

Some GMO crops have been designed to produce a natural protective pesticide-like by-product which is also detrimental to human health. Tilapia farming, done correctly, can be organic and about growing nutrient dense environmentally friendly food. GMO food and fish feed are contrary to logical reasoning, in regard to good health, concern for the environment, nutrient density, and optimal food production. Educated and responsible citizens try to not do anything that further promotes the GMO industry or proliferates GMO plants.

Being GMO-Free

There are several ways to avoid genetically altered foods:
- Purchase only organic foods.
- Reduce your consumption of processed foods, including store bakery goods wherever possible.
- Avoid artificial sweeteners, especially aspartame.
- Keep in mind that vitamin supplements may also contain GMO soy or corn ingredients. It is prudent to purchase only supplements that are 3rd party lab tested and noted as being non-GMO.
- While genetically modified animals are not currently approved for humans to eat, many GMO crops are used to feed animals that are later consumed by humans. The concentration of heavy metals and the toxins increase as they move up the food chain; with humans being at the end of the food chain. Unfortunately, many of these chemicals and heavy metals are not easily removed from the body after they are consumed—becoming significantly concentrated in us—leading to a higher propensity to health problems.
- To avoid eating animal products containing GMOs, purchase only organic, wild, or 100 percent grass-fed animals or those that were only fed organic feed.

There is sufficient reason to be concerned about GMO effects on health and the environment. Buying organic and arming yourself with the knowledge necessary to make informed choices, will help you avoid potential health risks to you and those you care about. In addition, every time you buy an organic product, your dollars serve as a voice to further support and strengthen the organic market. Being knowledgeable about GMOs and their negative impacts will also help you better market your produce.

Dirty Produce—Organic Substitute

The following list of produce (in no particular order) has been shown to have the greatest concentration of harmful pesticides and should only be consumed if grown organically. Since the pesticide, herbicide, and/or fungicide is often absorbed by the plant, simply washing the produce will not remove all of the toxic chemicals.

1. Peaches
2. Apples
3. Bell Peppers
4. Celery
5. Nectarines
6. Strawberries
7. Cherries
8. Kale / Lettuce /Spinach
9. Carrots
10. Pears
11. Cucumbers
12. Spinach
13. Blueberries
14. Grapes
15. Potatoes
16. Collard Greens

GMO Labeling

Chemical manufacturers, Big Agriculture and junk food companies spent more than $70 million spreading misinformation in 2012 to narrowly defeat GMO food labeling ballot initiatives in California and Washington. They continue to spend hundreds of millions of dollars in campaigning to defeat state GMO labeling laws in other states.

Currently, 64 countries have policies requiring the labeling of GMO foods, also referred to as Genetically Engineered foods, including all our trading partners in Asia, the European Union and even countries where Genetically Engineered (GE) crops are a large part of the economy, such as Brazil. While there may be universal agreement on labeling abroad, here in the U.S. multinational agribusinesses and food corporations spend millions of dollars in campaign contributions, lobbying efforts, and false advertising to block state labeling initiatives.

As a result of this outpouring of cash to defeat GMO labeling, Americans have been inundated with a number of myths intended to defeat state labeling efforts. These myths need to be corrected.

MYTH #1: **American consumers will see a huge spike in food prices if companies are required to disclose the GE ingredients that are already in their products.** The fact is companies change their labels all the time without causing a spike in the price of food. In fact, most companies do not print labels more than one year in advance for regulatory or marketing purposes. The establishment of a mandatory labeling standard for GE foods could easily fall within a company's regular label refresh cycle. The cost of modifying food labels is negligible.

MYTH #2: **Farmers are opposed to mandatory labeling.** While some may oppose labeling, major farm groups like the National Farmers Union, the National Family Farm Coalition and the National Black Farmers Association support mandatory labeling and have opposed legislation intended to block state laws in the absence of a national standard. With mandatory labeling there will be no new financial cost to farmers and given that food companies already produce foods for a variety of markets, there should not be additional segregation costs. Rather the information that is already being captured within the food supply chain will simply be provided to consumers on the end product.

MYTH #3: **We can rely on voluntary measures alone.** In the 14 years that the FDA has allowed companies to voluntarily label foods produced using genetic engineering, not one single company has done so. Similarly, we cannot merely rely on the use of voluntary absence claims like "GMO-Free." While such marketing claims allow companies to distinguish themselves in the marketplace, they are not a substitution for mandatory disclosure, because consumers are not given the full universe of information. Creating a federal standard for voluntary marketing claims will do nothing to address the overwhelming demand for labeling and likewise will do nothing to address consumer confusion that has festered in the absence of mandatory labeling.

MYTH #4: **We cannot label GE foods because there are too many and they are not dangerous.** In the U.S. we do not label dangerous food; we take it off the shelf. Rather, foods produced using genetic engineering are fundamentally different at the molecular and genetic level than those produced using conventional breeding methods. Mandatory labeling of GE foods is essential for preventing consumer deception and will allow con- sumers to make informed choices about the products they are buying and feeding their families.

FACT: **What is lost in these industry-driven myths (false advertisements) is the fundamental issue of equality.** Food issues can all too often turn into issues of class. Each and every person should have the right and the information available to know what they are buying and feeding to their families, regardless of where they shop and where they live.

It is time for Congress and all governments to provide leadership on an issue that impacts each and every person. Any costs would be negligible; and the benefits of labeling are numerous. Establishing a responsible national labeling standard that informs consumers of what is in the food they are buying and feeding their families is only right and fair.

Other Food Label Problems

Food labels are supposed to be there to help you make healthy, informed decisions about what you eat, in terms of not only the calories, sodium and fat content, but also the ingredients. However, a recent report released by the Government Accountability Office (GAO) gave the FDA a failing grade when it comes to preventing false and misleading labeling. The GAO also reported that the FDA does not track the correction of labeling violations, which means even if a food manufacturer is known to be using inaccurate labels, no one is checking up to make sure the problem is fixed.

Recently, "Good Morning America" hired a lab to test a dozen packaged food products to see if the nutrients matched the labels. All 12 products tested had label inaccuracies of some sort and three were actually off by more than 20 percent on items like sodium and total fat.

The US Government and GMOs

If you are wondering why the United States leads the world in GMO crop acreage, it's because the United States Department of Agriculture (USDA) and the FDA are heavily influenced by *Bayer* (previously known as Monsanto), which spends millions of dollars lobbying the U.S. government for favorable legislation that supports the spread of their toxic products every year. *Bayer* spends $1.4 to $2.5 million annually lobbying the federal government. Furthermore, they invest enormous sums of money on campaign contributions to politicians running for office that show favoritism towards their agenda. In addition, the U.S. Food and Drug Administration (FDA), the USDA, and the U.S. Trade Representative all have a special set of revolving doors leading straight to *Bayer*, which has allowed this transnational giant to gain phenomenal authority and influence. Former *Monsanto* and *Bayer* employees

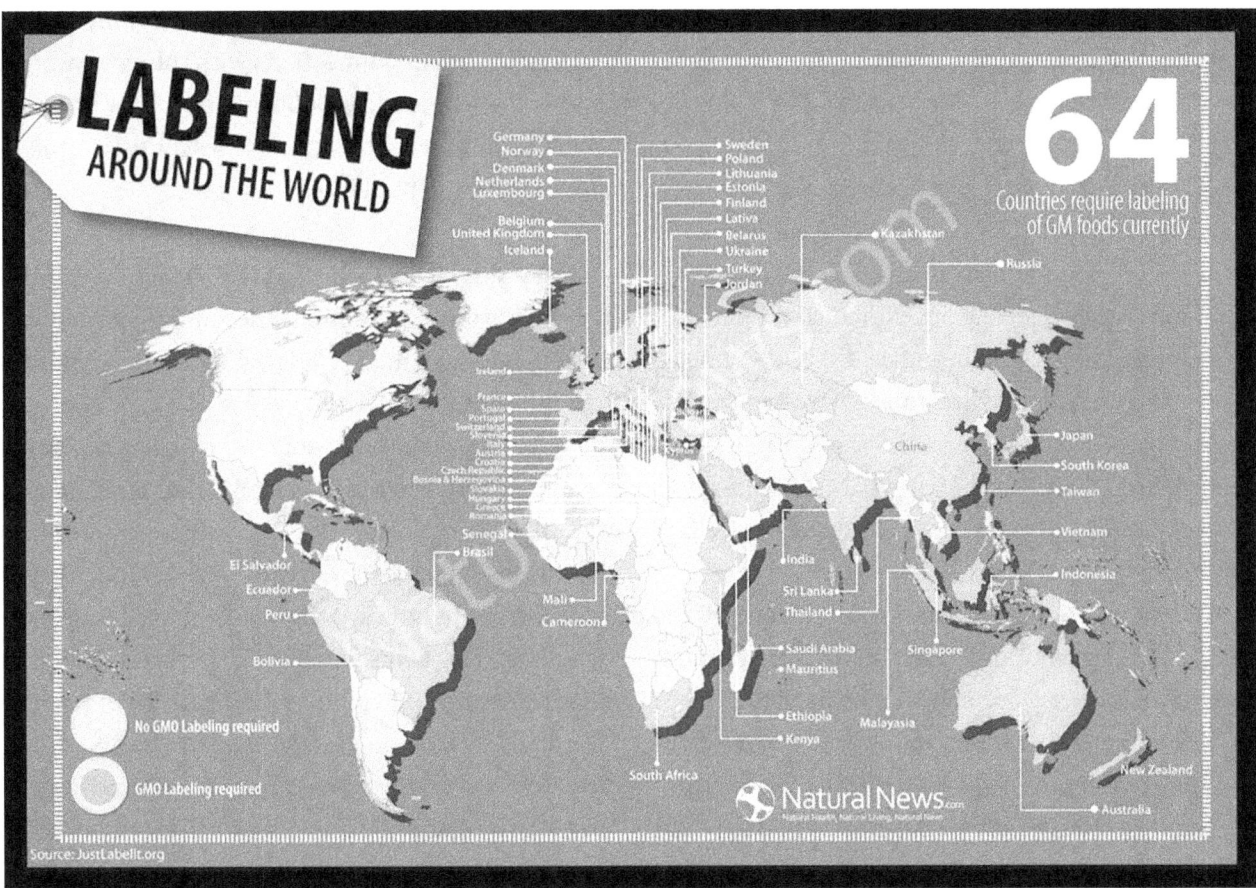

FIGURE 12. Worldwide GMO Labeling, GMOs are not labeled in the USA.

currently hold positions in US government agencies such as the Food and Drug Administration (FDA), United States Environmental Protection Agency (EPA) and the Supreme Court. Of few of these include:

- Clarence Thomas, Associate Justice of the Supreme Court.
- Michael R. Taylor, Deputy Commissioner for Foods at the USDA.
- Ann Veneman, former Executive Director of UNICEF, and US Secretary of Agriculture.
- Linda Fisher, Deputy Administrator of EPA.
- Michael Friedman, Deputy Director of the EPA.
- William D. Ruckelshaus, Administrator of EPA.
- Mickey Kantor, US Secretary of Commerce.
- Linda Fisher, Deputy Administrator of EPA. (Ms. Fisher has been back and forth between positions at Bayer and the EPA)

- Former Secretary of Defense Donald Rumsfeld was chairman and chief executive officer of G. D. Searle & Co., which Monsanto purchased in 1985. Rumsfeld personally made at least $12 million from the transaction.

The above list is just a brief summary of the more high-profile positions. Not listed are the hundreds of other GMO former company employees that are now employed in government positions. The cozy relationship our government has to GMO companies is also a two-way streak of opportunities. For instance, the GMO companies make it plain to government employees that wonderful future employment opportunities exist to those employees that are amicable to their agenda. Some of the more high-profile positions include:

- Josh King, former director of production for White house events, is now the director of global communication in Monsanto's Washington, D.C. Office.
- Clayton K. Yeutter, former Secretary of the USDA, former U.S. Trade representative who led U.S. negotiations in the U.S.-Canada Free Trade Agreement and helped launch the Uruguay round of the GATT negotiations, is now amember of the board of directors of Mycogen, whose majority owner is Dow. Mycogen is also the corporation thatholds the patent on a technology to genetically alter plants to produce and deliver "edible vaccines."
- Terry Medley, former administrator of the USDA Animal and Plant Health Inspection Serve, former chair and vice-chair of the USDA Biotechnology Council, and former member of the FDA Food Advisory Committee, is nowpresiding as the director of regulatory and external affairs of Dupont's agriculture enterprise.
- Micky Kantor, former Secretary of the US Dept. of Commerce and former US Trade Representative, is now a member of the board of directors of Bayer.
- Linda J. Fisher, a former Assistant Administrator of the EPA is now Vice-President of Public Affairs for Monsanto.
- William D. Ruckelshaus, the former chief administrator of the US EPA is now (and for the past 12 years) a memberof the board of directors of Bayer.
- Lidia Watrud, a former microbial biotechnology researcher at Bayer, is now with the US EPA.
- Margaret Miller, a former laboratory supervisor for Monsanto, is now Deputy Director of Human Food Safety andConsultative Services in the US FDA.

Again, the list above is a very brief summary showing just the more high-profile positions. There are many other employees that could be listed. Such shows a symbiotic culture where favoritism and rewards end up not only influencing but drive public policy. *Bayer* has similar connections with the government, and heavily influences public policy in thepharmaceutical and agricultural industries for their benefit, over that of the public.

Heavy Metals in Our Food Supply

Heavy metal poisoning is becoming a profoundly serious issue in America today, as toxins in groundwater and soils, some of which have persisted there for decades, are increasingly turning up in both drinking water and the general food supply.All across the country, residues of arsenic, lead and other heavy metals have been detected in well water, crop soils and even foods marketed toward the health conscious. Heavy metals are now being found in USDA certified organic foods, superfoods, vitamins, herbs, and dietary supplements at alarming levels. Neither the USDA nor the FDA have limited regulations on heavy metals in foods and organic foods, meaning that products can contain extremely toxic levels of mercury, lead, cadmium, arsenic, copper and even tungsten while still being legally sold across the USA.

The Natural News Forensic Food Lab, a very reputable independent laboratory headed by food researcher Mike Adams (AKA: the Health Ranger) has tested over 1,000 products using inductively coupled plasma mass spectrometry (ICP-MS) instrumentation. Below is a sample of some of the results obtained from the tests performed. A more comprehensive list of results is published at Labs.NaturalNews.com.

- Over 500 ppb Mercury in dried cat treats
- Over 10 ppm Tungsten in rice protein products
- Over 5 ppm Lead in ginkgo herb products
- Over 400 ppb Lead in cacao powders
- Over 500 ppb Lead and over 2000 ppb Cadmium in rice proteins
- Over 6 ppm Arsenic and over 1 ppm Lead in some spirulina products
- Over 100 ppb Mercury in dog treats

- Over 1,200 ppm Copper in children's multivitamins
- Over 200 ppb Lead in brand-name mascara products
- So much mercury contamination found in Maine lobster that the government put a temporary halt on applicable fisheries.

In response to the publishing of these findings some companies have said that they believe heavy metals are actually good for you and that their customers should eat more heavy metals; therefore, they are not going to make any efforts to reduce heavy metals in their products. This response stands at odds with all known environmental science and sound health advise.

The industry desperately needs a standard—even a voluntary standard—to which products can be compared for their heavy metals composition. Neither the USDA nor the FDA have expressed any interest in promoting or enforcingsuch a standard.

What Levels Are Safe for Consumption?

What is clear to nearly all environmental scientists are that lower exposure to dietary heavy metals is better for your health. All heavy metals interfere with healthy cellular function. At what level they become "dangerous" depends on your genetics, your diet, and your overall health.

Heavy metals bioaccumulate in the human body. When levels become high enough, they substantially interfere withhealthy functioning of the brain, liver, kidneys, heart, skin, reproductive organs, and other body systems.

- Learn more about heavy metals toxicity at http://labs.naturalnews.com/Heavy-Metals.html
- For expert clinical assistance regarding heavy metals detoxification, visit American College for Advancement in Medicine (ACAM) website at ACAM.org

Aren't Heavy Metals Safe Because They Naturally Occur in All Foods?

That is a myth promoted by some companies whose products are heavily contaminated with toxic heavy metals. Some of these companies are trying to convince consumers that heavy metals are not concerning by falsely claiming they are "naturally occurring." The claim is scientifically false, and a prime example of scientific fraud perpetrated against innocent ill-informed customers.

As proof of this, consider the fact that rice grown in China is often heavily contaminated with lead, cadmium, and mercury while rice grown in some areas of the USA is incredibly clean, with virtually zero levels of those same heavy metals. If heavy metals were "naturally occurring" in all rice, then levels would be the same no matter where rice is grown. However, this is not the case. This proves that much of the rice being grown in China is grown in areas which are heavily contaminated by industrial pollution containing heavy metals.

The Natural News Forensic Food Lab has also verified huge concentration differences in seaweed products, depending on where they are grown. Wakame grown near New Zealand has almost zero heavy metals, while Wakame grown near China is heavily contaminated.

Arsenic in Our Food Supply

Arsenic exists in both organic and inorganic forms. Both forms of arsenic are naturally occurring and are differentiated simply by the molecules in which the arsenic is attached. In quite simple terms, organic arsenic contains carbon and hydrogen, and inorganic arsenic contains other metals and elements, such as oxygen and sulfur. It has only been in the past few years that arsenic speciation (differentiation between organic and inorganic arsenic) in food has been possible.

Arsenic is the 55th most common element in the Earth's crust. Because arsenic is ubiquitous in the environment, it is unavoidable to have traces in

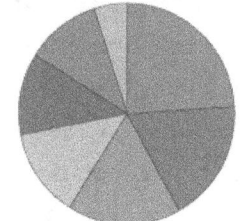

FIGURE 13.

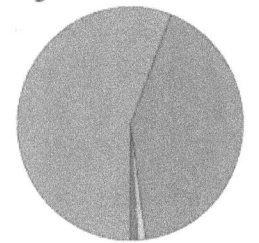

FIGURE 14.

our food supply given the obvious need for soil, air, and water to grow crops needed for direct human consumption and for feeding livestock. Arsenic is designated by the International Agency for Research on Cancer as a Class A human carcinogen.

Many public water supplies are also contaminated with arsenic because public utility filtration systems typically do not remove this toxin before delivering water to customers. This means that untold millions of people who drink unfiltered tap water are consuming arsenic daily.

Consumers Union, the public policy and advocacy arm of Consumer Reports, conducted laboratory tests of arsenic. Their testing, which included 200 samples of more than 60 different products, found inorganic arsenic at concerning levels in virtually every sample.

Exposure to Low-Level Arsenic is a Major Threat to Human Health

The U.S. Centers for Disease Control and Prevention (CDC) admits that even exposure to low levels of arsenic can cause major health damage over time. "Because it targets widely dispersed enzyme reactions, arsenic affects nearly all organ systems," says the CDC, noting that arsenic is also linked to gastrointestinal effects, renal damage, cardiovascular events, neurological damage, skin problems, anemia, leukemia, reproductive problems and cancer, including several types of skin cancer. "Arsenic can cause serious effects of the neurologic, respiratory, hematologic, cardiovascular, gastrointestinal, and other systems," adds the agency.

The Organic Trade Association research on arsenic revealed that the most commonly cited consequences of chronic exposure to low levels of arsenic include increased incidence of bladder, lung, kidney, and skin cancers, along with elevated levels of heart disease, skin hyperpigmentation, and skin lesions.

Despite it being a carcinogenic and causing so many health problems, there is no standard for arsenic in foods. There are no arsenic testing requirements. Furthermore, there are no regulations placed upon food companies which mandate labeling or warning the public of arsenic dangers in their food products.

Other Toxic Food Considerations

The primary goal is to achieve and maintain a healthy, happy life, devoid of the chronic illnesses and conditions plaguing society today. However, this has become increasingly difficult considering the numerous problems with the current food system in the United States. That is why growing our own food is so important.

The Food Industry

Understanding how the powerful $5+ Trillion-dollar food industry (expected to be $20 Trillion in 2030) works today is critically important for making the necessary decisions to achieve and maintain a healthy life. For example, approximately 90 percent of the money Americans spend on food is spent on 'processed' foods from large food manufacturers such as General Foods and Procter & Gamble. Their food marketers, employing some of the best and brightest minds to study consumer psychology and demographics, do a masterful job at targeting the right population and making it seem like their processed foods are the obvious, even healthy, best choice. But not only are these processed foods "dead" and devoid of any natural nutrition, they are often harmful to one's health and can even be loaded with carcinogenic substances.

The Media

Members of the media are subject to the same disinformation as the populace and therefore are often largely unaware of the problems in our current food system. A major funding source for the media, especially network television advertising revenue, comes from food and drug companies, making the information more unreliable, thus, either promoting unhealthy and dangerous products, or failing to report on the issues with industry. Celebrities and professional athletes often endorse products or participate in advertisements for processed food that they themselves may never or rarely consume, making the products seem more alluring.

Politicians and the FDA

Like the media, our elected officials are consumers who are subject to the same disinformation and are too often unaware of the health issues. Additionally, our politicians have been effectively controlled by the food and drug companies for so long that our government is now a large part of the problem, rather than being poised to be part of a solution. The Food and Drug Administration (FDA), although originally designed to protect consumers from unhealthy products, now often work diligently to protect the very companies it is supposed to regulate by keeping out competition and prolonging the economic life of the drug companies' government-sanctioned patents.

Sadly, government policy is now typically based on data supplied by the food or pharmaceutical industry, lobbyist, and campaign contributions rather than an science from and unbiased source. *USA TODAY* study found that more than half of the experts hired to advise the government on the safety and effectiveness of medicine and food have financial relationships with the companies that will be helped or hurt by their decisions. These experts are hired to advise the FDA on which substances should be approved for sale, and how studies should be designed. To date, the FDA has allowed more than 70,000 chemicals to infiltrate our food supply. Many of these dangerous ingredients are outright toxins, poisoning us, undermining our health, and allowing cancer and diseases to enter our bodies.

Fish

Fish is high in protein, and full of essential nutrients and healthy fats. It is also easily digested, and assimilated, and full of essential nutrients and fats. Unfortunately, the vast majority of fish sold is imported from developing countries, much of which have proven to be contaminated with banned chemicals, poisons, carcinogens, and high levels of antibiotics, often caused by being farmed or caught in contaminated waters.

However, imported fish is not the only fish that is likely to be bad for your health. According to a new U.S. Geological Survey study, scientists detected mercury contamination in every fish sampled in nearly 300 streams across the United States. More than a quarter of these fish were found to contain mercury at levels exceeding the criterion for the protection of people who consume average amounts of fish, established by the U.S. Environmental Protection Agency.

Exposure to mercury can damage your brain, kidney, and lungs. Mercury is especially damaging to your central nervous system (CNS), and studies show that mercury in the CNS causes psychological, neurological, and immunological problems. Furthermore, mercury has an extremely long half-life in the human body that scientist believe is somewhere between 15 and 30 years. Additionally, non-dissipating medications, antibiotics, fungicides, pesticides, herbicides, untreated sewage, and other chemicals are also entering our waterways at alarming rates.

Conventional raised farmed fish often found in grocery stores may not be the answer either. Studies show that farm-raised fish contain more polychlorinated biphenyl and over ten times the amount of dioxin. Raising Tilapia on feed that does not contain toxic chemicals is the way to go, as there are no dangerous substances entering the food supply.

Soy

Unfortunately, numerous so-called experts and the media have extolled soy a healthy food. Supporters claim it as an ideal source of protein. It is also being recommended as a great tool to for lowering cholesterol, protecting against cancer and heart disease, reducing menopause symptoms, and preventing osteoporosis, as well as other things. However, studies based on sound research have found that soy products may increase the risk of breast cancer, cause brain cause damage, contribute to thyroid disorders, promote kidney stones, weaken the immune system and cause fatal food allergies.

One of the most disturbing of soy's ill effects on health has to do with its phytoestrogens that can mimic the effects of the female hormone estrogen. Just moderate consumption of soy has been shown to alter a woman's menstrual cycle. This hormonal disruption is not healthy for men or women. Especially devastating for men, soy is known to lower testosterone levels.

Processed Meat

Processed meats are not a healthful choice for anyone and should be avoided entirely, according to a recent review of more than 7,000 clinical studies examining the connection between diet and cancer. The report was commissioned by The World Cancer Research Fund (WCRF). Virtually all processed meats contain a well-known carcinogen: sodium nitrite. It is a commonly used preservative and antimicrobial agent that also adds color and flavor to processed and cured meats, but this additive can also cause the formation of nitrosamines in your system, which can lead to cancer. Hot dogs, deli meats and bacon are notorious for their nitrite content.

Processed meat is far from being 100 percent beef, besides being loaded with harmful nitrates, it contains some nasty filler and by-product materials, such as other animal parts that most people find repulsive. It is also common for processed meats to contain MSG, high-fructose corn syrup, preservatives, and harmful artificial flavoring or artificial colors.

Processed Food

When discussing processed food, the list of problems is lengthy, but a few specific issues should be discussed. The first is the contaminants in processed food. The FDA, in its long history of managing food safety, has established guidelines for acceptable quantities of a number of contaminants that it will allow in our food supply. Included on the list are various types of mold, flies, maggots, insect eggs or body fragments, rodent hairs and feces pellets, pus pockets, and larvae, to name just a few. No, there is not a typo, the FDA has established acceptable limits of these contaminates.

Another issue with processed foods is the hydrogenated oils used to lengthen the shelf life of products like crackers and cookies, which are also associated with diabetes and heart disease. They are also generally high in sodium, corn syrup and other unhealthy ingredients.

As with charred meats, researchers discovered that a cancer-causing and potentially neurotoxic chemical called acrylamide is created when carbohydrate-rich foods are cooked at high temperatures, whether baked, fried, roasted, grilled, or toasted. The chemical is formed from a reaction between sugars and an amino acid (asparagine) during high-temperature cooking above 212°F. As a general rule, the chemical is formed when food is heated enough to produce a fairly dry and "browned" surface. Hence, it can be found in processed potatoes, grain products and even coffee.

Acrylamide is not the only hazard associated with heat-processed foods, however. The three-year long EU project known as *Heat-Generated Food Toxicants*1 *(HEA*TOX), identified more than 800 heat-induced compounds in food, 52 of which are potential carcinogens. For example, the high heat of grilling reacts with proteins in red meat, poultry, and fish, creating heterocyclic amines, which have also been linked to cancer.

Food Additives

More than 3,000 food additives, such as preservatives, flavorings, colors, and other ingredients, are legally added to foods in the United States, most of which are unhealthy and often dangerous. Many of these additives have been linked to an increased risk of cancer, while recent studies have shown others are estrogen-mimicking xenoestrogens that have been linked to a range of hormonal health effects in males and females. Studies have also shown that a variety of common food dyes and preservatives cause some people to become measurably more hyperactive and distractible and can even do brain damage resulting in a significant reduction in IQ.

International Foods (China)

Generally speaking, quality control and employee health are not issues that the Chinese food industry or the Chinese government are overly concerned about. Without the regulation, China's products often contain plastics, pesticides, herbicides, and other cancer-causing chemicals, in far greater amounts than those produced in America.

The United States imported roughly 4.6 billion pounds of agricultural products from China in 2014. For example, China is responsible for 90 percent of the vitamin (C) consumed by Americans, 78 percent of the tilapia, 70 percent of the apple juice, 50 percent of the cod, 43 percent of the processed mushrooms and 23 percent of the garlic. The top U.S. import commodities from China are fruits and vegetables (fresh/processed), snack food, spices, and tea – the combined which accounts for nearly one-half of the total U.S. agricultural imports from China. According to congressional testimony by Don Kraemer, Deputy Director of the Office of Food Safety at the FDA, *"The FDA has encountered compliance problems with several Chinese food exports, including lead and cadmium in ceramic-ware used to store and ship food, and staphylococcal contamination of products from China remains a concern for FDA, Congress, and American consumers."* **Less than 1 percent of food imports are examined (laboratory tested) in any given year!**

Artificial Sweeteners

All Artificial Sweeteners are toxic to the human body. They basically trick your body into thinking that it is going to receive sugar (calories), but when the sugar doesn't come your body continues to signal that it needs more, which results in carb cravings. Contrary to industry claims, research over the last 30 years—including several large-scale prospective cohort studies—has shown that artificial sweeteners stimulate appetite, increase cravings for carbs, and produce a variety of metabolic dysfunctions that promote fat storage and weight gain.

The most common artificial sweeteners are Aspartame *(NutraSweet, Equal)*, Acesulfame-K or Ace-K *(Sunett, SweetOne)*, Saccharin *(Sweet'n Low)*, and Sucralose *(Splenda)*. The list of side effects includes, but are not limited to headaches, chronic respiratory disease, dermatologic reactions, tachycardia, cancer, and tumors, enlarged organs, and many neuro-psychiatric disorders, including panic attacks, mood changes, visual hallucinations, manic episodes, isolated dizziness, memory impairment, nausea, temper outbursts, and depression.

Stevia isn't really an artificial sweetener because it is made from one of the extracts of the stevia plant (an herb); and is often marketed as a natural sweetener. However, *Stevia* can negatively interfere with absorption of carbohydrates and can further disrupt the metabolizing and conversion of food into energy.

Food Colorants

Americans are now eating 5 times more food dye than in 1955. There are serious hidden dangers of food colorants. Commonly, they can cause various tumors, allergic reactions, brain gliomas, immune system impairment, and hyperactivity. With their widespread use, they can be found in baked goods, beverages, candies, cereals, treated fruit, ice cream, cosmetics and even dog food. So if you are concerned about your health you should avoid foods with these common label ingredients: Blue #1, Blue #2, Citrus Red #2, Green #3, Red #40, Red #3, Yellow #5 (Tartrazine), Yellow #6 or any other food dyes.

Preservatives

Preservatives are chemicals used to keep food fresh. Although there are a number of different types of food preservatives, antimicrobials, antioxidants, and products that slow the natural ripening process are some of the most common. Despite their function, preservatives can pose a number of serious health risks.

Cancer is a serious side effect associated with the use of preservatives. In fact, the National Toxicology Program reports that propyl gallate—a preservative commonly used to stabilize certain cosmetics and foods containing fat—may cause tumors in the brain, thyroid, and pancreas. Similarly, *InChem*—an organization that provides peer-reviewed information on chemicals and contaminants—notes that nitrosamines, including nitrates and nitrites, can lead to the development of certain cancer-causing compounds as they interact with natural stomach acids. Nitrosamines are found in a variety of foods, including cured meat, beer, and non-fat dried milk.

Hyperactivity in children is another possible side effect associated with the use of preservatives. A study performed by the *Archives of Disease in Childhood* (ADC) noted a significant increase in hyperactive behavior in 3-year-olds who took benzoate preservatives. Children who were enrolled in the study also demonstrated a decrease in hyperactive behavior after they stopped taking benzoate preservatives. While benzoates can be found in a number of foods, they are often used to preserve acidic foods and beverages, like soda, pickles, and fruit juice.

Some people may experience damage to their heart as a result of preservative use. Sodium nitrates can cause blood vessels to narrow and become stiffer. In addition, nitrates may affect the way the body processes sugar and may be to blame for the development of some types of diabetes, the Harvard School of Public Health notes.

Studies have shown and even the FDA has warned that other preservatives, such as Chlorphenesin and Phenoxyethanol, sulfites, BHA (Butylated hydroxyanisole) and BHT (Butylated hydroxytoluene) are toxic and in addition to the above-named effects, can cause depression of the central nervous system, vomiting and diarrhea, severe allergic reactions including anaphylactic shock, stomach irritation, skin sensitivity, and even effects on the body's blood coagulation system.

Flavor Enhancers

Flavor enhancers are used in foods to enhance the existing flavor or modify flavors in the food without contributing to any significant flavor of their own. Food flavor enhancers are commercially produced in the form of instant soups, frozen dinners, and most processed foods. Monosodium glutamate (MSG) is one of the most common flavor enhancers. The FDA has indicated flavor enhancers ingested in low doses are generally regarded as safe (GRAS); however, the list of dangerous side effects is extensive. Most are known to be carcinogenic to humans, but can also cause brain damage, neurological diseases, kidney problems, obesity, eye damage, headaches, depression, fatigue, disorientation, chest pain or difficulty breathing, and severe allergies.

High-Fructose Corn Syrup (HFCS)

Food companies like to use high-fructose syrup, instead of traditional sweeteners because it costs less to make, is sweeter to the taste, and mixes more easily with other ingredients. The average American consumes nearly 63 pounds of high-fructose syrup a year. High-fructose corn syrup is commonly added to many processed foods and beverages. It maybe be listed on food labels as "corn sweetener," "corn syrup," or "corn syrup solids" as well as "high-fructose corn syrup".

Research shows that this liquid sweetener can upset the human metabolism, raising the risk for heart disease and diabetes. Researchers say that high-fructose corn syrup's chemical structure also encourages overeating. It forces the liver to pump more heart-threatening triglycerides into the bloodstream. In addition, it can deplete the body's reserves of chromium, a mineral important for healthy levels of cholesterol, insulin, and blood sugar. Additionally, almost half of tested samples of commercial HFCS contained mercury. Mercury was also found in nearly a third of 55 popular brand-name food and beverage products where HFCS is the first- or second-highest labeled ingredient. Mercury is toxic in all forms, and a poison to the brain and nervous system. At a minimum it causes extreme fatigue and neuro-muscular dysfunction

Food Labels

Food companies know that people are drawn to certain terms like "All Natural" and "Made with Whole Grains." They can make enormous profits if they can market their products as being healthy. "Natural" labeled food generated $219 billion in 2019, and 60 percent of all cereals are now labeled "whole grain," even processed, sugar-laden ones. Misleading food labels regularly dupe consumers with these keywords, questionable advertising practices, misleading packaging statements, and bold statements that appeal to people's dietary preferences and weight loss goals. Sometimes it is impossible to identify the ingredients in a product by reading the label. The FDA allows for ingredients that are present onlyin small quantities to be labeled as "artificial flavor", "natural flavor", or "spices". Only the manufacturer knows what all the ingredients are in the product. Unfortunately, since the FDA does not regulate all food labels it is unable to keep food manufacturers from using crafty wording.

Antibiotics in Food

Antibiotics are widely used in food-producing animals and is now viewed as a normal operating practice. Animals in factory farms are given hefty doses of antibiotics so that they can remain alive in stressful, unsanitary conditions, and to make them grow faster. Even many grazing animals, not raised in farm lots, are given antibiotics to help them grow faster. The practice of giving animals antibiotics greatly increases profit margins for livestock producers and pharmaceutical companies. As a result, according to the FDA more than 32 million pounds of antibiotics are given to animals annually.

This use contributes to the emergence of antibiotic-resistant bacteria in food-producing animals. These resistant bacteria can contaminate the foods that come from those animals, as these antibiotics work their way up the food chain, and persons who consume these foods can develop antibiotic-resistant infections. Scientists around the world have provided strong evidence that antibiotic use in food-producing animals can have a negative impact on public health. Because of the link between antibiotic use in food-producing animals and the occurrence of antibiotic-resistant infections in humans, CDC encourages and supports efforts to minimize inappropriate use of antibiotics in humans and animals.

In addition to the emergence of antibiotic-resistant bacteria, common side effects from antibiotics include diarrhea, nausea, vomiting, fungal infections, and a variety of allergic reactions.

Fluoride/Chromium-6/Chlorine and Bottled Water (Water Treatments)

Fluoridation is not legal or not used in the overwhelming number of countries including industrialized ones. In fact, 97 percent of Western Europe has chosen fluoride-free water. The United States is one of only eight countries in the entire developed world that fluoridates its water supply. It is added under the guise that it prevents and control tooth decay.

The fluoride added to our drinking water is actually chemical waste product. It is certainly not anything that should be ingested or included in our diet. The chemicals used to fluoridate water in the US are not pharmaceutical grade. Instead, they come from the wet scrubbing systems of the superphosphate fertilizer industry. These chemicals (90 per-cent of which are sodium fluorosilicate and fluorosilicic acid), are classified hazardous wastes and contaminated with various impurities.

Numerous studies have shown that fluoride is dangerous to our health. Five studies from China show a lowering of IQ in children associated with fluoride exposure. The Department of Health in New Jersey found that bone cancer in male children was far greater in areas where water was fluoridated. The U.S. Environmental Protection Agency (EPA) researchers confirmed bone cancer-causing effects of fluoride at low levels in their animal studies. Sadly, 23 human studies and 100 animal studies have linked fluoride to brain damage. Even the US Food and Drug Administration (FDA) has never approved any fluoride product designed for ingestion as safe or effective. Although they still allow it to be put into our drinking water, food, and toothpaste. Read the warning label on your toothpaste about fluoride, then throw it in the trash and start using fluoride free toothpaste and mouthwash.

Fluoride is dangerous, but Chromium-6 (also known as hexavalent chromium) in your water supply may be far worse. Chromium-6 was detected in 31 of the 35 city water supplies tested and confirmed by a number of independent tests by various water utility companies. Hexavalent chromium is classified as "likely to be carcinogenic to humans" by the U.S. Environmental Protection Agency (EPA).

Chlorine is another noxious chemical deliberately put in the water by public health officials. Chlorine is in the same chemical group as fluoride, which has been linked with cancer and osteoporosis. The chemical element chlorine is a corrosive, poisonous, greenish-yellow gas that has a suffocating odor and is 2-1/2 times heavier than air. It can resultin arteriosclerosis, heart attack and stroke. Research has shown that individuals who consume chlorinated drinking water have an elevated risk of cancer of the bladder, stomach, pancreas, kidney, and rectum as well as Hodgkin's and non-Hodgkin's lymphoma.

Bottled water has issues, as well. A recent study on more than 250 bottles from 11 leading brands worldwide revealed that a single liter of bottled water can contain dozens or even tens of thousands of microplastic particles, which is much greater than the levels found in tap water samples. Microplastics hold toxic chemicals that can harmfully impact human health, the wildlife, and natural environments.

The best way to ensure that you are drinking healthy water is to install a reputable filtration system. Reverse osmosisfilter systems are some of the strongest, most effective filters for drinking water. They are known to remove more than 99% of most dangerous contaminants in the water, including heavy metals, herbicides, pesticides, chlorine, and other chemicals, and even hormones. A good system ranges in cost from $250 to $600 (USD 2020) and can be installed at the kitchen sink. Healthy fish tank water will be addressed later.

Pesticides/Herbicides

As discussed earlier in this chapter pesticides are widely used in producing food to control pests such as insects, rodents, weeds, bacteria, mold, and fungus. Seven of the most toxic chemical compounds known to man are approved for use as pesticides in the production of foods. These toxins are referred to as Persistent Organic Pollutants (POP's). They are called persistent because they are not easily removed from the environment. The greatest risk to our environment and our health comes from the chemical pesticides. In spite of the dangers, the government maintains its approval of the use of toxic chemicals to make pesticides. And science is constantly developing variations of poisons.

Pesticides can be toxic to humans and animals. It can take a small amount of some toxins to kill, and other toxins that are slower acting may take a long time to cause harm to the human body. Even just using pesticides in amounts within regulation, studies have revealed neurotoxins can do serious damage. Several factors determine how your body will react including your level of exposure, the type of chemical you ingest, and your individual resistance to the chemicals. Some possible reactions are fatigue, skin irritations, brain and blood disorders, nausea and vomiting, liver and kidney damage, reproductive damage, breathing problems, cancer and even death.

Herbicides (weed killers) are mixtures of chemicals designed to spray on weeds, where they get inside the plants and inhibit enzymes required for the plant to live. The active ingredient in the most widely used herbicide is glyphosate, and some have the main component of Agent Orange. Until the introduction of GMO crops about 20 years ago, herbicides were sprayed on fields before planting, and then only sparingly used around crops. The food that we ate from the plants was free of these chemicals.

In stark contrast, with herbicide resistant GMO plants, the herbicides, and a mixture of other chemicals (surfactants) required to get the active ingredient into the plant are sprayed directly on the crops and are then taken up into the plant. The surrounding weeds are killed while the GMO plant is engineered to resist the herbicide. Therefore, the food crop itself contains the herbicide as well as a mixture of surfactants.

To accommodate the fact that weeds are becoming glyphosate resistant, thereby requiring more herbicide use, the EPA has steadily increased its allowable concentration limit in food and has essentially ignored our exposure to the other chemicals that are in its commercial formulation.

As a result, the amount of glyphosate-based herbicide introduced into our foods has increased enormously since the introduction of GMO crops. Multiple studies have shown that glyphosate-based herbicides are toxic and likely public health hazards.

New scientific studies link glyphosate to a host of health risks, such as cancer, miscarriages, and disruption of human sex hormones. Other conditions with strong correlations are ADHD, Alzheimer's disease, anencephaly, autism, birth defects, brain, breast and other cancers, celiac disease, chronic kidney disease, colitis, depression, diabetes, heart disease, hypothyroidism, inflammatory bowel disease, liver disease, and many others.

Irradiation

Irradiation is the process of exposing fresh foods to low amounts of x-rays to sterilize and prolong its life. Commercialized food irradiation is done on a constant basis. Food growers and sellers say that food irradiation is safe. However, studies show the x-ray irradiation destroys vitamins A, E, and K. Some studies show that x-ray irradiation also destroys up to 85 percent of the vitamin C found in vegetables, and fruits. Irradiation also destroy the digestive enzymes in raw foods. Irradiation damages food by breaking up molecules and creating free radicals. Free radicals are unstable atoms that damage cells, causing illness, accelerate the aging process, and contribute to a host of diseases, such as cancer. When high-energy electron beams are used, trace amounts of radioactivity may be created in the food. Some bacteria, like the one that causes botulism, aswell as viruses are not killed by current doses of irradiation.

No one knows the long-term effects of a life-long diet that includes foods which will be frequently irradiated, such as meat, chicken, vegetables, fruits, salads, sprouts, and juices. Studies on animals fed irradiated foods have shown increased tumors, reproductive failures, and kidney damage. Some possible causes are irradiation-induced vitamin deficiencies, the inactivity of enzymes in the food, DNA damage, and toxic radiolytic products in the food.

Bisphenol-A (BPA)

BPA is one of the biggest players in the wrapping industry. In 2017, an estimated six million tons of BPA-derived chemicals were produced, making it one of the highest produced chemicals worldwide. In addition, BPA has an expected compound annual growth rate of 6% over the period 2019–2024. BPA is deeply imbedded in the products of modern consumer society, not just as the building block for polycarbonate plastic but also in the manufacture of epoxy resins and other plastics. BPA is routinely used to in plastic containers, water bottles, to line cans to prevent corrosion and food contamination. BPA has been used as an inert ingredient in pesticides, as a fungicide, antioxidant, flame retardant, rubber chemical, and polyvinyl chloride stabilizer. BPA contamination has become widespread in the environment in rivers and estuaries throughout the US. Unfortunately, BPA does not readily degrade in the environment. What this all means is that we are constantly being exposed to BPA. A CDC (Centers for Disease Control and Prevention) study found 95 percent of adult human urine samples and 93 percent of samples in children had BPA. However, the FDA has approved the use of BPA and the EPA does not consider it cause for concern.

According to estimates, just a couple of servings of canned food can exceed the daily safety limits for BPA exposure in children. Even low-level exposure to BPA can be hazardous to one's health. There are more than 100 independent studies linking the chemical to serious health problems in humans. Some of the main problems are prostate cancer and breast cancer, diabetes, and obesity, altered immune function, early sexual development in girls and disrupted reproductive function, learning and behavioral problems, including hyperactivity, abnormal heart rhythms and coronary artery disease, asthma, depression, diabetes, heart disease and reproductive disorders.

Growth Hormones

Today's hyper-productive animals are given injections and implants (in the case of cows) or genetic engineering (in the case of salmon), of artificially high levels of sex or growth hormones. This allows them to grow bigger, faster and produce better. Surprisingly, little research has been done on the health effects of these hormones in humans, in part because it's difficult to separate the effects of added hormones from the mixture of natural hormones, proteins, and other components found in milk and meat.

However, many experts have valid concerns that these excess hormones in the food supply are contributing to cancer, early puberty in girls, and other health problems in humans. For years, consumer advocates and public health experts have fought to limit the use of hormones in cows, and some support a ban on the practice. Unfortunately, the FDA continues to approve of growth hormones in animals for human consumption, even though there has been minimal testing to determine whether or not there are any health consequences related to their use.

In 1993, the FDA approved recombinant bovine growth hormone (rBGH), a synthetic cow hormone that increases milk production when injected into dairy cows. Research has found that milk from rBGH-treated cows contains up to 10 times more IGF than other milk. Higher blood levels of IGF have been associated with more than a 50 percent increased risk of breast, prostate, and other cancers in humans.

IGF isn't the only hormone found in the food supply. Ranchers have been fattening up cattle with other sex hormones, most notably estrogen, since the 1950s. Today most beef cows in the U.S, except those, that are raised 'organic', receive an implant in their ear that delivers a hormone, usually a form of estrogen (estradiol) in some combination with five other hormones. Even miniscule amounts of estrogen could affect prepubescent girls and boys. Whether it is related or not, sperm counts in men from America, Europe, Australia, and New Zealand have dropped by more than 50 percent in less than 40 years. The majority of public health experts, health-conscious nutritionist, and most registered dietitians urge consumers to stay away from rBGH-treated milk because of its potentially higher IGF levels.

Biomagnification & Bioaccumulation

Biomagnification, also known as **bioamplification** or **biological magnification**, is the increasing concentration of a substance, such as a toxic chemical, in the tissues of organisms at successively higher levels in a food chain. This increase can occur as a result of:

- **Persistence** – where the substance cannot be broken down by environmental processes.
- **Food chain energetics** – where the substance's concentration increases progressively as it moves up a food chain.
- **Low or non-existent rate of internal degradation or excretion of the substance** – often due to water-insolubility.

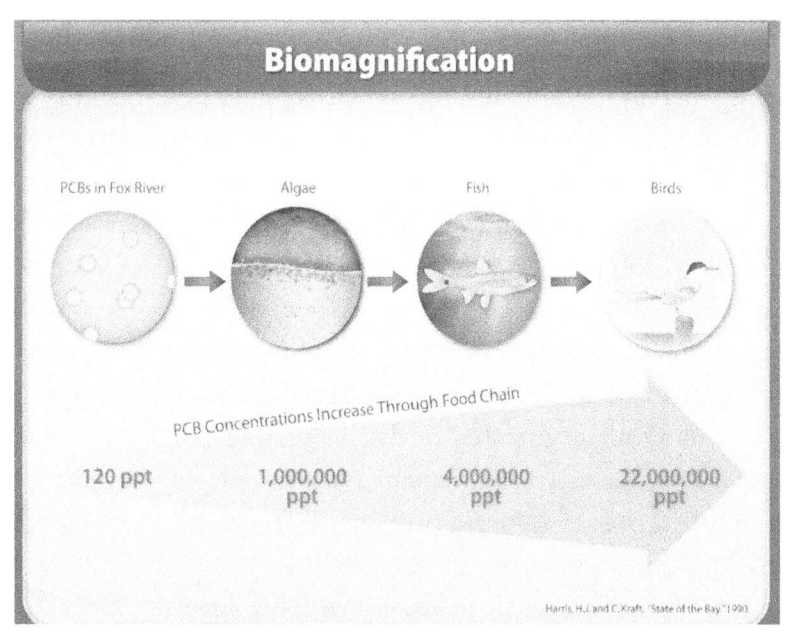

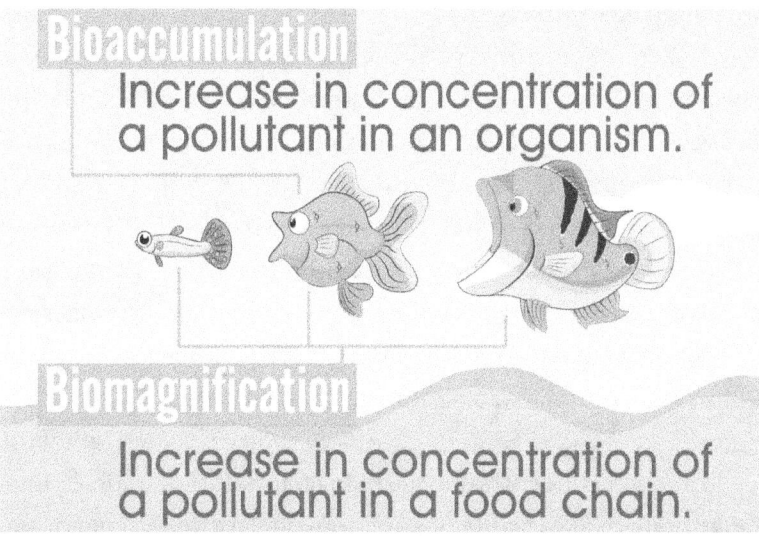

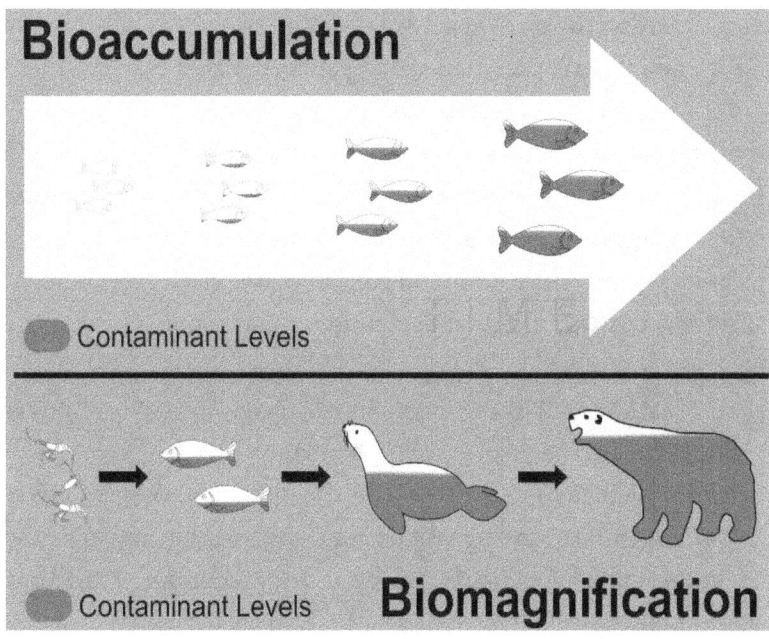

Biological magnification often refers to the process whereby certain substances such as pesticides or heavy metals work their way into lakes, rivers, and the ocean, and then move up the food chain in progressively greater concentrations as they are incorporated into the diet of aquatic organisms such as zooplankton, which in turn are eaten perhaps by fish, which then may be eaten by bigger fish, large birds, animals, or humans. The substances become increasingly concentrated in tissues or internal organs as they move up the chain. Bioaccumulates are substances that increase in concentration in living organisms as they take in contaminated air, water, or food because the substances are very slowly metabolized or excreted.

Although sometimes used interchangeably with "bioaccumulation", an important distinction is drawn between the two, and with bioconcentration.

- **Bioaccumulation** – the increase in the concentration of a substance in certain tissues of organisms' bodies due to absorption from food and the environment.
- **Bioconcentration** is defined as occurring when uptake from the water is greater than excretion.

Thus, bioconcentration and bioaccumulation occur within an organism, and biomagnification occurs across trophic (food chain) levels.

Dolphins have been studied by Japanese researchers as a model species for biomagnification because their migratory routes are known, they live in relatively unpolluted waters, and they live a long time (20-50 years). DDT has been found in dolphin blubber in greater concentrations (100 times greater than sardines) than would be expected given the small concentrations present in the water and in sardines, their favorite food. These unexpectedly large concentrations are the result of DDT biomagnification up the food chain.

Biomagnification has serious consequences for all species. It is particularly dangerous for predator species especially if they are at the top of long food chains. Predators are usually at or near the top of their food chain. This puts them at risk because the degree of biomagnification is high by the time it reaches their trophic level. Also, top predators usually consume large quantities of meat which has lots of fatty tissue and contaminants. Polar bears, humans, eagles, and dolphins are examples of top predators, and all of these organisms are vulnerable to the effects of biomagnification. Predators that consume large amounts of wild caught fish have an exceptionally high degree of risk.

Medical Industry vs. Nutrition

Nutrition and diet have long been considered the basis for health. People in the 18th century knew scurvy could only be cured and/or prevented by foods in the diet. So, in today's very advanced age of technology and research, it stands to reason medical professionals would be well-versed in the effects of nutrition on their patients' health. Unfortunately, this is not the case.

Numerous sources report that doctors themselves feel they are inadequately trained to discuss nutrition with their patients. An article titled "A Time for Change: Nutrition Education in Medicine" by nutrition.org states,

> "...among a cohort of 930 cardiologists, 90% believe their role includes providing patients with basic nutrition information. In the same group of physicians, though, 90% stated that they had received little-to-no training in nutrition during their fellowship, 59% stated that they had received no nutrition during internal medicine training, and 31% reported no nutrition education in medical school."

Nutrition.org also reported results from a recent survey which indicated 71% of medical schools "failed to meet the minimum nutrition education recommendation of 25 hours, 36% provided 12 or fewer hours, and 9% provided none." Why are medical professionals not being given the training/education they need to advise their patients in this integral part of their health?

One doctor has been quoted as saying, "The problem is that there is not really a consensus on how important nutrition is, and what kind of nutrition should be taught." Dr. Rupy Aujla in his article "Food is medicine – so why aren't our doctors trained in the science of nutrition?" states, "If we don't appropriately educate our health professionals on the breadth and utility of other evidence-based health interventions, the only option they will have heard of is a pharmaceutical one." The pharmaceutical industry is capitalizing on this opportunity through intense marketing strategies that begin when students are in medical school. Rijul Kshirsagar and Priscilla Vu in their article "The Pharmaceutical Industry's Role in U.S. Medical Education" states, "Medical students' exposure to pharmaceutical marketing begins early, growing in frequency throughout their training. Students receive gifts such as free meals, textbooks, pocket texts, small trinkets and even drug samples. Pharmaceutical companies, recognizing the formative nature of the clinical years of medical education, seek to form relationships with medical students years before they are ready to independently practice medicine." Therefore, with little to no nutrition training and constant exposure to pharmaceutical options, many doctors will prescribe a pill to address a symptom rather than discuss nutrition with a patient to address and potentially prevent a problem.

Additionally, physicians receive extraordinarily little education in prevention. As Thomas J. Van Gilder and Patrick Remington in their article "Medical education needs to take 'an ounce of prevention' seriously" states "Traditional medical education may include a token course in prevention, but the focus is squarely on how to treat patients."

While many health issues such as obesity, diabetes, congestive heart failure, and other conditions are preventable, the ability to prevent them requires physician knowledge about nutrition and/or a willingness to partner with other professionals knowledgeable about nutrition to form a care team. Unfortunately, it appears there is little funding, training and/or incentives to support preventive care; and the medical community is rewarded greatly though tremendous monetary gains for implementing a pharmaceutical treatment plan.

Overweight & Obesity

Approximately 1/3 of the world's population is either obese or overweight. Obesity is an issue affecting people of all ages and incomes, everywhere. The rise in global obesity rates over the last three decades has been substantial and widespread, presenting a major public health epidemic in both the developed and the developing world.

Worldwide obesity has nearly tripled since 1975. More than 50% of the world's 671 million obese live in 10 countries (ranked beginning with the countries with the most obese people): US, China, India, Russia, Brazil, Mexico, Egypt, Germany, Pakistan, and Indonesia. The Middle East showed large increases in obesity. Bahrain, Egypt, Saudi Arabia, Oman, and Kuwait were among the countries with the largest increases in obesity globally. In six countries, all in the Middle East and Oceania – Kuwait, Kiribati, the Federated States of Micronesia, Libya, Qatar, and Samoa – the prevalence of obesity for women exceeds 50%. In Tonga, both men and women have obesity prevalence over 50%. In sub-Saharan

Africa, the highest obesity rates (42%) are seen among South African women. In the United States, more than 71% of all adults over the age of 20 are either overweight or obese.

Overweight and obesity results in substantial health risks -- such as cardiovascular disease,

cancer, diabetes, osteoarthritis, dementia, and chronic kidney disease increase substantially when a person is overweight. Children who are overweight or obese are at greater risk for high blood pressure, type 2 diabetes, and heart disease. According to the National Institutes of Health, obesity and overweight together are the second leading cause of preventable death in the United States, close behind tobacco use.

Overweight and obesity is a financial issue. For instance, the obesity crisis costs the United States more than $150 billion in healthcare costs annually and billions of dollars more in lost productivity.

Overweight and obesity is a national security issue. The obesity crisis impacts the United States military readiness. Being overweight or obese is the leading cause of medical disqualifications, with nearly one-quarter of service applicants rejected for exceeding the weight or body fat standards. Obese service members and members of their family who are obese cost the military about $1 billion every year in healthcare costs and lost productivity.

Overweight and obesity is a community safety issue. With millions of obese and overweight Americans serving as first responders, firefighters, police officers and in other essential community service and protection roles, public safety is at risk. Seventy percent of firefighters are overweight or obese, putting them at risk for cardiovascular events — the leading cause of line-of-duty deaths.

Overweight and obesity is a child development and academic achievement issue. Childhood obesity is correlated with poor educational performance and increased risk for bullying and depression.

Overweight and obesity is a top national priority. Americans rated obesity as the top health concern in the country in a recent public opinion survey conducted by the Greenberg, Quinlan, Rosner Research and Bellwether Research groups.

The main driver of obesity is NOT lack of exercise. For years we have been told that a lack of exercise and a sedentary lifestyle are the major factors associated with the obesity crises. However, physical activity levels began to decline before the global obesity rate started to surge — which means changes to the food environment are the prime obesity culprit. Food companies, convenience stores, and fast-food restaurants have made inroads all over the world with cheap, calorie-dense, low-nutrient content food. Soda, candy, processed food, and fast food is loaded with preservatives, additives, sugar, sodium, and other harmful substances. These products are inexpensive, heavily marketed, and are much more accessible than healthier alternatives, like fruits and vegetables. In addition, portion sizes have gone up and people are eating outside of the home more often, which means heavier unhealthy meals.

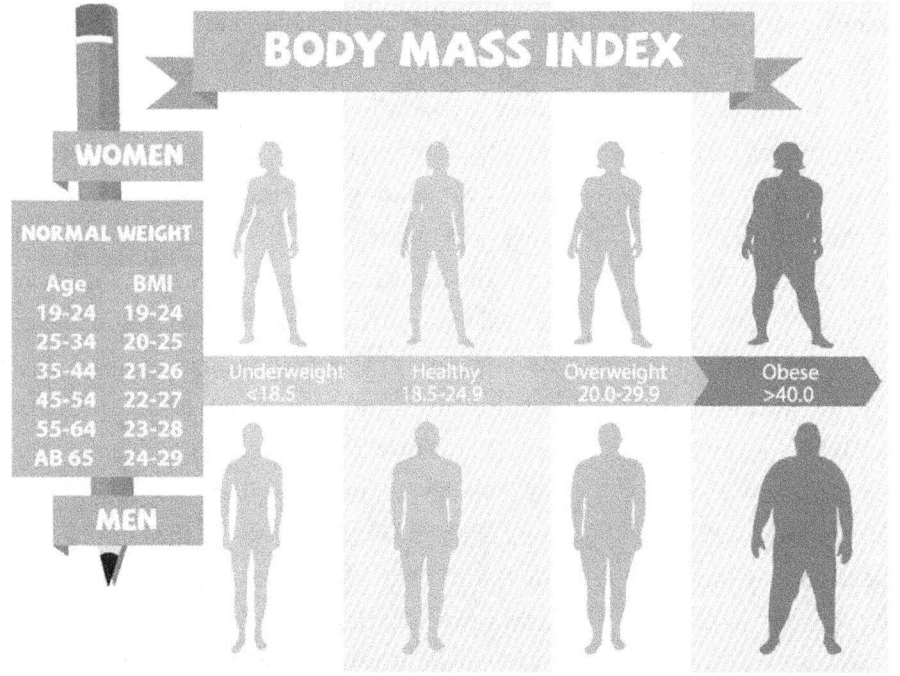

FIGURE 15. NOTE: To determine your BMI enter "BMI Calculator" in your favorite online search engine.

The government (influenced by millions of dollars in food industry lobbyist efforts and political campaign contributions) and the media (influenced by millions of dollars in food industry advertisements) tell us we need to exercise more as the way to shrink our waistlines. The exercise chant for weight loss is highly misleading. Exercise is excellent for health; and should be an important part of everyone's day, as should proper stress management, staying hydrated, getting adequate sleep, stretching daily, and avoiding things that are harmful to your body. Although exercise can aid in weight loss and weight control, it is not nearly as important as eating properly on a consistent basis. Eating healthy foods, avoid junk foods and processed foods, implementing portion control, and managing caloric intake are the keys to weight loss and weight management.

What really works for weight loss is not just cutting calories but satisfying your hunger with the right kinds of foods. Eating a diet of healthy fish, fruits, and vegetables, instead of fast food and processed food will do more for weight loss than anything else. In addition, a person on this type of diet feels better and will enjoy a higher quality of life. Coincidently, Tilapia farming provides healthy fish, and exercise.

The Solution

You may be asking yourself why a book about farming tilapia has gone into so much detail about tour food industry, toxic food, and the associated problems they cause. There are several reasons why I think it is important that you be armed with this knowledge.

First, it will help you to make better healthy food choices for yourself and your loved ones. Healthier food choices result in a longer, more productive, and better quality of life.

Secondly, understanding the dangers associated with society's food system reinforces the importance of farming tilapia in healthy ways, doing aquaponics, and organic gardening. Properly growing our own food lowers the risk that we and our loved ones are exposed to unhealthy toxic foods.

Lastly, being knowledgeable about the dangers of our toxic food system will better enable you to market, barter, and sell your healthy grown product (e.g., tilapia, fruits, vegetables).

Properly Raised Tilapia -- Healthy for Humans

Tilapia is an inexpensive, mild-flavored fish. It is the fourth most commonly consumed type of seafood in the United States, after shrimp, tuna, and salmon. Many people love tilapia because it is relatively affordable and doesn't taste very fishy.

Tilapia is an Excellent Source of Protein and Nutrients

Tilapia is a pretty impressive source of protein. In 3.5 ounces (100 grams), it packs 26 grams of protein and only 128 calories. Even more impressive is the amount of vitamins and minerals in this fish. Tilapia is rich in niacin, vitamin B12, phosphorus, selenium, and potassium. A 3.5-ounce serving contains the following:

- Calories: 128
- Carbs: 0 grams
- Protein: 26 grams
- Fats: 3 grams
- Niacin: 24% of the RDI
- Vitamin B12: 31% of the RDI
- Phosphorus: 20% of the RDI
- Selenium: 78% of the RDI
- Potassium: 20% of the RDI

Tilapia is also a lean source of protein, with only 3 grams of fat per serving.

Another thing to keep in mind, especially if you are looking for farm-raised fish fed with non-GMO feed: The USDA does not currently have guidelines for classifying seafood as organic.

CHAPTER 3

Location and Setup Considerations

To achieve maximum success with any type of aquaculture operation it needs to be setup and managed utilizing optimal location criteria conditions. Setup and location are extremely important considerations that cannot be ignored when farming tilapia.

Fish Parameters

Fish prefer to be in the shade. A fish tank should never be under direct sunlight, as it can cause increased temperatures and promote algae blooms in the tank. Algae blooms can cause a lack of oxygen which, in turn, kill the fish. Placing a tank under a roof or covering will address this issue. When raising tilapia in a pond, growing floating plants on the surface of the water will provide, shade, shelter, and hiding places for the fish. Hiding places make fish feel more secure and amounts to less stress on them.

Space Allowance and Accessibility

It is also especially important to have good accessibility to the fish tank for feeding, checking water quality and the health of the fish, as well as harvesting the fish. Since fish tanks are heavy, it is essential that they be properly supported or positioned on solid flooring (e.g., secure ground, concrete, etc.). Elevating fish tanks can be costly and complicated.

Power Considerations

An aquaculture system needs to have access to power for the pump in order to keep the water circulating. Lighting is not always required, but certainly beneficial. Power for temperature management is also a consideration in some environments. Power needs for lighting and temperature management is dependent upon the particular system set-up, location, and the operator's goals.

Consideration should also be given to for the possibility of long-term power outages due to a storm event, or other infrastructure failure. It is prudent to have a readily available secondary power source, such as a generator or an inverter box connected to an automobile battery, available in the event the power goes out for an extended time. This will ensure the life supporting equipment (e.g., pump, oxygen diffuser, etc.) can continue operating and fish kill risk is minimized.

Some operators prefer to have their aquaculture system off the grid. As a result, they obtain their electrical needs system via solar or wind power.

Alternative power systems are covered in detail later in this book.

Heating and Insulation

Different fish species have their own optimal temperature ranges, but they all tend to prefer uniform water temperatures. Depending on the location of the aquaculture operation, it may be necessary to have a submersible heater (which requires electricity) to keep the water temperature within the optimal range. Some operators use heat wire around their water line(s).

Heating is not required if the fish being raised has an optimal temperature range within the same temperature to achieve maximum fish growth and ideal beneficial bacteria conditions. However, a heater may provide some advantages, even if the average ambient air temperature is usually in the life support range. Hence, even though the cost of electricity is higher, there can be increased yields with the addition of a heater, which results in greater production yields. If there is a potential for significant temperature fluctuation as a result of location, then insulation around the tank and submersible heaters to keep the water temperature constant and within optimal range is necessary; otherwise, the growth of fish can be stunted, or they may even die.

Ergonomically Correct

It is also helpful to plan a system that is ergonomically correct. In other words, top of the fish tank, equipment and devices should be designed to fit the human body and its correct functional abilities. Such will maximize productivity by reducing operator fatigue and discomfort. An ergonomically correct layout will not only be much more convenient and help optimize production, but it will make the entire aquaculture endeavor a much more enjoyable experience.

If the top of the fish tank has to be high, then consider building a platform to stand on. The fish tank can also be partially imbedded in the ground so that the top working surface is at a lower user-friendly height. Step ladders come in handy for tall tanks and to reach lights. However, the system should be designed so that the use of a step ladder is not a regular requirement.

Also, keep in mind that walking surfaces should be safe, free of tripping hazards, and have a non-slip surface when wet. The design height of all aquaculture components should be based on a chart, or architectural standard (as will be presented in the system design chapters later in this book), but an even better approach is to construct the system specifically to the operator's body height— that being what is most convenient for the end user.

Storage Spaces

There are a lot of materials, tools, and supplies (piping, tubes, fish food, etc.) needed in order to keep an aquaculture system running smoothly. Easy accessible storage spaces nearby comes in very handy, as there will be frequent trips to this area.

Indoor Operations

One of the benefits of having the system indoors is that the fish tank will serve as a wonderful source of thermal heating in the winter and thermal cooling in the warmer months. However, there is no hard and fast rule as to how the system must be set up. In some parts of the world everything can be located outside.

Expansion and Transitioning into a Profitable Business

Ideally, when planning your aquaculture system your designed layout will be such that it provides you the flexibility to grow as you gain experience and are ready to scale up. It is easier to manage, operate, trouble-shoot problems when things go wrong with a smaller system compared to that of a large operation. This is especially important when you are still relatively new to aquaculture. Furthermore, any errors with a smaller system typically do not hurt as much in regards to cost and time. However, mistakes can serve as wonderful opportunities to learn and improve, so they should not be viewed entirely as a bad thing, as they can be excellent teachers. Problems in a smaller system are just easier to recover from. Therefore, starting small and growing as you learn is usually the best approach. Just keep system expansion in mind when you are planning your design layout. Designating and preserving an area for future expansion, during your initial planning phase, will save you a tremendous amount of trouble, expense, and time later, compared to the complexity of relocating or rearranging an existing system.

If your desire is to transition into a profitable business operation, keep in mind that profit typically progresses with experience. The more you learn, the more you will earn.

FIGURE 16. *Seachem Multi-Test Nitrite and Nitrate Test Kit*

Health Check of Fish

Unhealthy fish are often a warning that the system is out of balance. If fish exhibit signs of stress (gasping at the surface; rubbing on the sides of the tank; showing red areas around the fins, eyes, and gills; or, in extreme cases, dying), it is often because of a buildup of toxic ammonia or nitrite levels. This often happens when there is too much dissolved waste in the pond or for the media in the filter process in a tank. Obviously, any of these symptoms in the fish or plants indicate that the operator needs to actively investigate and rectify the cause.

Nitrogen Testing

This method involves testing the nitrogen levels in the water using simple and inexpensive water test kits (example shown below). If ammonia or nitrite are high >1 ppm (>1 mg/liter), the filter is inadequate and should be increased. Most fish are intolerant of these levels for more than a few days. Fish can typically tolerate short-term elevated levels of nitrate, but if the levels remain high >150 ppm (>150 mg/liter) for several weeks some of the water should be removed and replaced.

- 75 tests can be performed with this kit.
- Cost: $18 (USA, year 2021).
- Cost and abilities of water test kits vary considerably. The below kit, or similar from another manufacturer, is sufficient for aquaponic operations.

Be sure to review Chapter 15 'Water Quality' for additional information and helpful resources.

PART II

Fish

CHAPTER 4

Fish — Everything You Need to Know

Importance of Fish

Fish serve as a valuable source of low-fat, high quality protein, essential omega-3 fatty acids, vitamin D, riboflavin, and other necessary nutrients. Growing a large population of fish may be a little daunting to some, especially those without any prior experience; however, one need not be discouraged. Successfully raising fish is just a matter of following some basic fundamental principles. As you follow these simple, proven principles, you will not have any difficulty growing fish from fingerling size to ready-to-eat fish. Chapter 16 "Fish Breeding, Fish Reproduction, and Raising Your Own Crop of Fish" will explain in detail how you can easily breed and raise your own fish, so as to spare you the cost of having to purchase fingerlings.

Tilapia for Profitability

To generate the maximum revenue, consider the following issues:

- **Choose a marketable species.** A good example of a species that is easy to produce, but can be difficult to market, is common carp. It is advisable to consider other, more widely accepted species. Of the sixty or so potential fish species used for food, channel catfish, **tilapia**, crawfish, rainbow

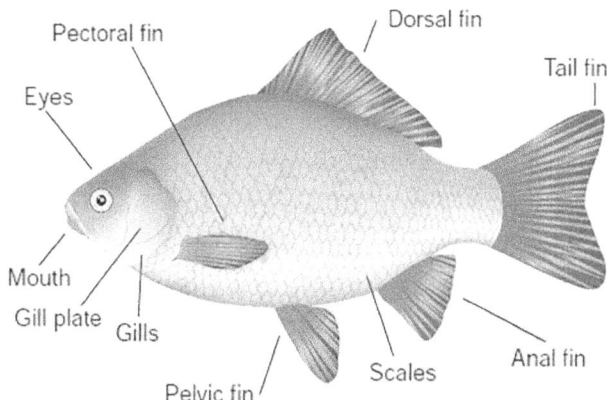

FIGURE 17. *Illustration of the main external anatomical features of fish*

trout, and salmon have large, established industries in the United States.

- **Know the complete production cycle.** Without complete production information, raising fish becomes a very risky venture.
- **Raise a species that is in high demand.** When choosing what fish to rear, you will need to consider what types of fish are in high demand in the market where your species you like to consume that cost you the most at the store (or which would be most profitable to sell) and ensure that the preferred type of fish is suitable to your climate. Fortunately, tilapia farming done in a healthy way checks all the boxes.

Tilapia Fish Details

In the United States, tilapia is a species that is extremely popular in aquaponics systems, for beginners, and in backyard aquaculture operations. They are an ideal species for these types of arrangements for many reasons. They are easy to breed, fast growing, can withstand poor water conditions, consume an omnivorous diet, and are a delicious edible fish. They are very resilient as well, being less sensitive than other fish species to fluctuations in temperature, pH, dissolved oxygen levels, and a build-up in waste. They can also grow well in crowded tanks. Tilapias have especially high beneficial nutritional ratios of omega-6 to omega-3 fatty acids. Tilapias grow to plate size in about 6-9 months.

A potential downfall is that they may not be the most ideal species for systems located in cooler climates as they require warm water. Every breed of tilapia is different, but most tilapia prefer temperatures of 77 to 86°F (25-30°C). They start to lose their resistance to disease and infections below 54°F (12°C). If you live in a cooler area and want to avoid large heating costs, you are much better off growing a fish species that will do well in your temperature range. However, if you are growing tilapia in a climate-controlled environment (e.g., warehouse, greenhouse, heated garage, etc.) then living in a colder climate is not an issue if the space is heated anyway.

Tilapias are a declared an invasive species in some areas of the United States. Therefore, they are prohibited by U.S. Fish and Game laws, and/or other governing agencies in some regions of the country, even if raised in indoor tanks (e.g., northern California).

There are over 100 different species of tilapia, each with unique characteristics and behavior. Members of the Oreochromis genus are a common choice for aquaponics or aquaculture. In terms of popularity, the Nile tilapia (*O. niloticus*) is the most widely cultured tilapia, followed by Blue tilapia (*O. aureus*), and Mozambique tilapia (*O. mossambicus*). All can be reproduced fairly easily by a beginner.

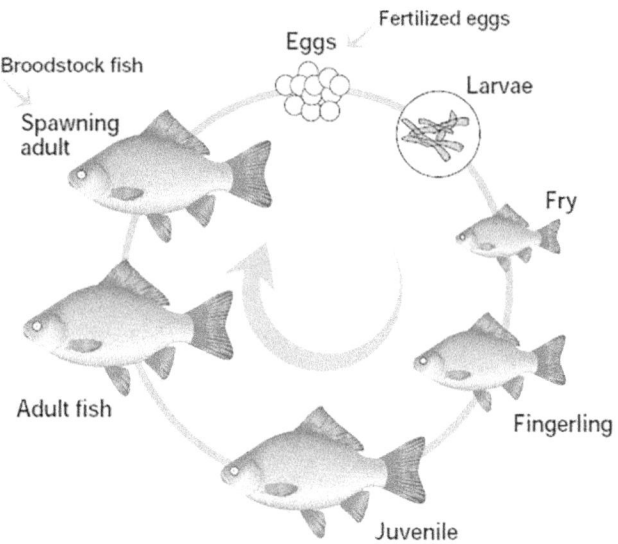

FIGURE 18. *General life cycle of a fish*

Tilapias start out as omnivores, but later become more like vegetarians. The tilapia's digestive system is designed to eat algae, vegetation, other small fish, worms, and insects. The newly hatched, called 'fry', require a lot of protein for fast growth. You can actually stunt their growth by under feeding them, which is what some breeders do to keep their fish in the sellable fingerling size range. As they mature, they require less protein and are not interested in eating their own fry, which is why it is possible to raise fry with adult tilapia in a single tank. Be sure to find a high-quality tilapia feed that takes all this under consideration. Be diligent in checking out the protein and fat contents in the fish feed formula when shopping to ensure that it provides the necessary nutrients for each stage of the tilapia's growth.

Tilapias are often fed pellet fish food, duckweed, and green plant products. Some operators grow algae and duckweed in separate tanks to feed their Tilapia.

Commercial tilapia farms will usually feed their fish pellets made from fishmeal, grain, soybeans, or other food products. In the wild, tilapia will eat vegetation, algae, plankton, insects, larvae, decaying organic matter, fish wastes, small fish and just about anything edible that they can get in their mouth. As stated previously, tilapia can be vegetarians and survive just fine by eating only algae.

Tilapia can be cultured as mixed-sexes (males and females together) or as mono-sex (males only). Most large-scale commercial growers prefer to raise only male tilapia because they grow faster than the females. Male tilapia populations can be produced by visual selection, hybridization, sex-reversal, and genetic manipulation. Tilapia with a minimum weight of 25 to 30 g can be separated by visual inspection of the genital papilla.

During the last 10 to 15 years, the most popular way to produce all-male populations is with hormone sex reversal of tilapia fry. Recently hatched tilapia fry obtained by harvest from spawning containers 18 days after brood fish are stocked or hatched from eggs taken from females are fed a powdered diet containing a male steroid for 20 to 28 days. Fry that would have been females if fed a steroid-free diet, will be functional males at the end of the hormone treatment. While all-male populations are hard to produce with sex reversal treatment, 95 to 98 percent males are commonly produced. Tilapia reach sexual maturity at 3 to 5 months of age.

The three methods of raising fingerlings including the open pond, hapa, and tank method. Hapas are net or cage enclosures placed in ponds to help contain the tilapia and keep predators out

Recommended Tilapia

SCIENTIFIC NAME: Oreochromis niloticus, O. aurea, O. mossambicus, O. hornorum

PRODUCTION POTENTIAL: Easy except tilapia grow slowly at temperatures lower than 70°F and die when temperatures drop into the 50°F range.

MARKET SIZE: 1.25–2.0 lbs

MARKET: Food

TEMPERATURE REQUIREMENTS: *Growing*: 80–87°F, *Spawning*: Greater than 72°F, Lethal: 55°F

FEED REQUIREMENTS: *Protein*: 25–30%, *Fat*: 6–8%

SPAWNING REQUIREMENTS: Maternal mouth brooders, spawn twice a month, 2,500 eggs/lb body weight, eggs hatch in 5-7 days. All male hybrids can be produced by crossing female O. niloticus and male O. aurea. or by crossing female O. niloticus and male O. hornorum. Stocking ratios for fingerling production is three females to each male.

Tilapia Feed Conversion Ratio Feeding Rates, Growth Rate, and Harvesting

The feed conversion ratio (FCR) or feed conversion efficiency (FCE), to define it simply, is a measure of an animal's efficiency in converting feed mass into increased body mass. Tilapias have an average feed

TABLE 1. **Tilapia Growth and Feeding Rates**

MONTH	START WEIGHT (g)	END WEIGHT (g)	GROWTH RATE (g/day)	FEEDING RATE (% weight)
1	1	5	0.2	15–10
2	5	20	0.5	10–7
3	20	50	1.0	7–4
4	50	100	1.5	4–3.5
5	100	165	2.0	3.5–2.5
6	165	250	2.5	2.5–1.5
7	250	350	3.0	1.5–1.25
8	350	475	4.0	1.25–1
9	475	625	5.0	1

conversion rate of 1.7. This 1.7 to 1 ration is incredibly efficient, especially compared to cattle which have a FCR of 8:1 (8-lbs of feed for every 1-lb of mass).

It takes between 7 and 9 months for a tilapia to reach harvest size of 1.25-lbs (0.57-kg or 567-grams) which produces two, 4-ounce (113-grams) fillets, or enough for a meal for two people. Commercial growers, using supplemental feedings and intensive management techniques, are able to achieve harvest of a 500-gram fish within five months. Keep in mind, though, that premium prices for your fish will only be achieved via being able to honestly market your fish as being grown through sustainable farming practices, fully organic (no artificial supplements and organic based feed products), in an animal friendly environment.

Tilapia can be harvested anytime they reach a marketable size of 1 to 2 pounds. Tilapia grow fast and usually reach market size in 6 to 12 months, depending on the feed and conditions. Large commercial operators generally harvest all of their fish at one time, whereas smaller operators typically harvest their fish over a period of time. Both options have their advantages. Harvesting all at once is more efficient, and the fish usually are sold to one buyer (i.e., a food company). For the small to mid-size operator, harvesting over time relieves system overcrowding, minimizes water quality challenges, and proves a steady supply of fresh fish for meals and/or an ongoing revenue stream.

Tilapias are most commonly processed into skinless, boneless fillets. The fillet yield is typically 30 to 37 percent of their total body weight, depending on fillet size and final trim. [1-ounce = 28.3 grams]

Fish Overview — Common Aquaculture Edible Fish

Nearly all freshwater fish are edible, but their taste, production potential, and growing parameters must be taken into consideration when raising them. It is vital to effectively manage all parameters associated with the fish to have a successful operation. Managing water quality and using only top-rated fish food helps ensure that your fish stay thrive and remain healthy.

Another thing to consider is whether to buy fry or fingerlings. Fish fry is cheaper, but they take longer to mature. On the other hand, fingerlings are more expensive, but they can be harvested sooner.

When they have developed to the point where they are capable of feeding themselves, the fish are called fry. When, in addition, they have developed scales and working fins, the transition to a juvenile fish is complete and it is called a fingerling. Fry are fish just after they are hatched. They are very small and often actually look like dirt in the water. Fry are

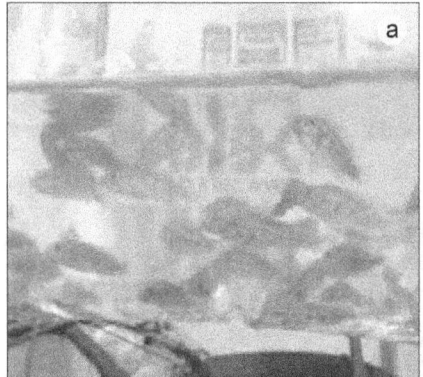  

FIGURE 19. Acclimatizing fish. Juvenile fish are transported in a plastic bag (a) which is floated in the receiving tank (b) and the fish are released (c).

TABLE 2. **Tilapia Life Support Parameters**

LIFE SUPPORT	TILAPIA
Thriving Temp.	74°–80°F
	24°–26°C
Surviving Temp.	60°–95°F
	16°–35°C
Carnivore or Omnivore	Omnivore
Oxygen Needs	Low

stocked in ponds during the early spring as soon as the eggs hatch.

Fingerlings vary in size but are typically about the size of fingers. Fingerlings are stocked in ponds during the summer and fall when their chances of survival are greater.

Acclimatizing Fish

Acclimatizing fish into new tanks or pond can be a highly stressful process for fish, particularly the actual transport from one location to another in bags or small tanks. It is important to try to remove as many stressful factors as possible that can cause fatality in new fish. There are two main factors that cause stress when acclimatizing fish: changes in temperature and pH between the original water and new water; therefore, differences in pH and temperature from the previous fish environment to the new tank must be kept to a minimum for the fish to have the best chance at survival during the transfer.

The pH of the culture water and transport water should ideally be tested. If the pH values are more than 0.5 different, then the fish will need at least 24 hours to adjust. It is ideal to keep the fish in a small, aerated tank in their original water if possible, and slowly add water from the new tank over the course of a day. Even if the pH values of the two environments are fairly close, the fish still need to acclimatize. The best method to do this is to slowly allow the temperature to equilibrate by floating the sealed transportation bags containing the fish in the culture water. This should be done for at least 15 minutes. At this time small amounts of water should be added from the culture water to the transport water with the fish. Again, this should take at least 15 minutes, so as to slowly acclimatize the fish. Larger volumes of transfer water require longer water temperature adjustment times. Finally, the fish can be added to the new tank.

Stocking Density

The recommended stocking density is 1 pound of fish per 5–7 gallons of tank water (0.5 kg per 20-26 liters). However, a lighter fish stock density has been found to be more forgiving if things go wrong (e.g., a pump or filter malfunction, etc.).

Keep in mind that there are other factors affecting the number of fish that can be grown in an aquaculture system, such as the species of fish and feed rates. The more the fish are fed, the more waste they produce. It is important to monitor system parameters such as water flowrates, oxygen levels, pump rates, and water temperature as they all play a critical role as well. However, don't let these things intimidate or discourage you from getting into aquaculture. This book will provide you the information needed so that you may be successful.

Harvesting and Staggered Fish Stock

A constant biomass of tilapia in the tank, cage, or pond makes feeding filtering easier. To achieve a constant biomass a staggered stocking method should be adopted. This technique involves maintaining three age classes, or cohorts, within the same tank, cage, or pond. Approximately every three months, the mature fish (500 g each) are harvested and immediately restocked with newfingerlings (50 g each). This method avoids harvesting all the fish at once, and instead retains a more consistent biomass. Tables 3 and 4, outline the potential growth of tilapia over one year using a progressive harvesting technique. The important aspect of this table is that the total weight of the fish varies between

TABLE 3. **Potential growth rates of tilapia in one tank over a year using a progressive harvest technique**

MONTH	DEC	JAN	FEB	MAR	APR	MAY	JUN	JUL	AUG	SEP	OCT	NOV	DEC
STOCKING ROUND	WEIGHT (KG)												
1	1.5	3.75	6.0	8.25	10.5	12.75	15.0*						
2				1.5	3.75	6.0	8.25	10.5	12.75	15.0*			
3							1.5	3.75	6.0	8.25	10.5	12.75	15.0*
4										1.5	3.75	6	8.25
5													1.5
Total fish mass (kg)	1.5	3.75	6.0	9.75	14.25	18.75	24.75–9.75	14.25	18.75	24.75–9.75	14.25	18.75	24.75–9.75
Action							Restock harvest			Restock harvest			Restock harvest

NOTES: Fingerling tilapia (1.5 kg = 50 g/fish x 30 fish) are stocked every three months. Each fish survives and grows to harvest size (15g = 500 g/fish x 30 fish) in six months. The asterisk indicates harvest. The range during harvest/stocking months accounts for the range if not all 30 fish are taken at once, i.e. the 30 mature fish are harvested throughout the month. This table serves only as a theoretical guide to illustrate staggered harvest and stocking in ideal conditions.

TABLE 4. **Potential growth rates of tilapia in one tank over a year using a progressive harvest technique**

MONTH	DEC	JAN	FEB	MAR	APR	MAY	JUN	JUL	AUG	SEP	OCT	NOV	DEC
STOCKING ROUND 1													
Number of fish in tank	80	80	70	60	50	40	30	10					
Fish weight (g)	50	125	200	275	350	425	500	575					
Cohort biomass (kg)	4	10	14	17	18	17	15	5.8					
STOCKING ROUND 2													
Number of fish in tank							80	80	70	60	50	40	30
Fish weight (g)							50	125	200	275	350	425	500
Cohort biomass (kg)							4	10	14	17	18	17	15
Total tank biomass (kg)	4	10	14	17	18	17	19	15.8	14	17	18	17	15

NOTES: Tilapia fingerling are stocked every six months. Staggered harvest starts from the third month to keep the total fish below the maximum stocking biomass of 20 kg/m³. The table shows the theoretical weight of each batch of harvested fish along the year if fish are reared in ideal conditions.

10–25 kg, with an average biomass of 17 kg. This table is a basic guideline depicting optimum conditions for fish growth. In reality factors such as water temperature and stressful environments for fish can distort the figures presented here.

If it is not possible to obtain fingerlings regularly, a system can be still managed by stocking a higher number of juvenile fish and by progressively harvesting them during the season to maintain a stable biomass. The first harvest starts from the third month onward. Various combinations in stocking frequency, fish number, and weight can apply providing that fish biomass stands below the maximum limit of 20 kg/m3. If the fish are mixed-sex, the harvest must firstly target the females to avoid breeding when they reach sexual maturity from the age of five months. Breeding depresses the whole cohort. In the case of mixed-sex tilapia, fish can be initially stocked in a cage and males can then be set free in the tank after sex determination.

It is important to note that adult tilapia will predate their smaller siblings if they are stocked together. A technique to keep all of these fish safely in the same fish tank is to isolate the smaller ones in a floating frame. This frame is essentially a floating cage, which can be constructed as a cube with PVC pipe used as frame and covered with plastic mesh. It is important to ensure that larger fish cannot enter the floating cage over the top, so make sure that the sides extend at least 6-inches (15 cm) above the water level. Each of the vulnerable size classes should be kept in separate floating frames in the main fish tank. As the fish grow large enough not to be in danger, they can be moved into the main tank. With this method, it is possible to have up to three different stocking weights in one tank, so it is important that the fish feed pellet size can be eaten by all sizes of fish. Caged fish also have the advantage of being closely monitored to determine the feed rate ratio by measuring the weight increment and weight of the feed over a period.

Important Notes Regarding Fish

- Feed the fish as much as they can eat in 5-10 minutes, two times per day. Remove uneaten feed in tanks after 30 minutes. Record total feed added. Balance the feeding rate with the number of plants using the feed rate ratio, but avoid over or under feeding the fish.
- Fish appetite is directly related to water temperature, particularly for topical fish such as tilapia, so remember to adjust feeding during colder winter months.
- A fingerling tilapia (50 g) will reach harvest size (500 g) in 6–8 weeks under ideal conditions. Staggered stocking is a technique which involves stocking a system with new fingerlings each time some of the mature fish are harvested.

Keeping Fish Healthy

An aquaculture system fish tank or pond has inputs of food, water, and oxygen. They have outputs of urine, ammonia, carbon dioxide, feces, and uneaten food.

Although each fish species is unique in regard to the ideal growing conditions, there are some generalities that are common for all fish. For instance, most fish need the pH to be somewhere between 6 and 8 (the optimal pH range for Tilapia is between 7 and 9).

Fish need oxygen (they will die within 30 minutes without it). Drastic temperature changes can cause health issues and even result in death. Fish are sensitive to light and do best when not exposed to direct light. Fish need to be temperature matched to the water before releasing them into the tank.

Ammonia and nitrites are very toxic to fish. Nitrates, on the other hand, are fairly safe for fish. Nitrates and nitrites are two different types of compound. Nitrates (NO_3) consist of one nitrogen atom and three oxygen atoms. Nitrites (NO_2) consist of one nitrogen atom and two oxygen atoms. Nitrates are relatively inert, which means they're stable and unlikely to change and cause harm.

Fish Maintenance

The following are general rules to keep in mind:
- Feed fish 2-3 times a day, but don't overfeed.
- Fish eat 1.5-2.0 percent of their body weight per day.
- Only feed fish what they can eat in 5-10 minutes.
- Fish won't eat if they are too cold, too hot, or stressed.
- Check water quality regularly.
- Add water or do partial water changes when necessary.
- In addition to checking water conditions of the fish tank, observe fish behavior and appearance. After becoming familiar with your fish, you will often be able to tell if there is something wrong or different.
- Some fish desire to be your friend and will regularly "greet you" at the tank.

TABLE 5. **Cause and symptoms of stress in fish**

CAUSES OF STRESS	SYMPTOMS OF STRESS
Temperature outside of range, or fast temperature changes	Poor appetite
pH outside of the 7 to 9 range or fast pH changes (more than 0.3/day)	Unusual swimming behavior, resting at surface or bottom
Ammonia, nitrite or toxins present in high levels	Rubbing or scraping the sides of the tank, piping at surface, red blotches and streaks
Dissolved oxygen is too low	Piping at surface
Malnourishment and/or overcrowding	Fins are clamped close to their body, physical injuries
Poor water quality	Fast breathing
Poor fish handling, noise or light disturbance	Erratic behavior
Bullying companions	Physical injuries

Types of Fish Disease

Fish ailments can be separated into four general types:
* Bacterial Diseases
* Fungal Diseases
* Parasitic Diseases
* Physical Ailments

How to Prevent Fish Diseases

Precautions can and should be taken to reduce the possibility of your fish getting a disease. Following these precautions can also help keep fish diseases from spreading if they do occur:
* Buy only good quality, compatible fish, and then breed/raise your own fish thereafter.
* Quarantine new fish in a separate tank before adding them to your aquaponic system.
* Avoid stressing the fish with rough handling, sudden changes in conditions, or "bully" tank mates.
* Do not overfeed your fish.
* Remove sick fish to a separate tank for treatment.
* Disinfect nets used to move sick fish.
* Do not transfer water from the quarantine tank to the main tank.
* Do not let any metal that has the propensity to rust come in contact with your system's water.

Recognizing Disease

Diseases may occur even with all of the prevention techniques listed above. It is important to stay vigilant and monitor and observe fish behavior daily to recognize the diseases early. The following lists outline common physical and behavioral symptoms of diseases.

External Signs of Disease:

* Ulcers on body surface, discolored patches, white or black spots,
* Ragged fins, exposed fin rays,
* Gill and fin necrosis and decay,
* Abnormal body configuration, twisted spine, deformed jaws,
* Extended abdomen, swollen appearance,
* Cotton-like lesions on the body,
* Swollen, popped-out eyes (exophthalmia).

Behavioral Signs of Disease:

* Poor appetite, changes in feeding habits,
* Lethargy, different swimming patterns, listlessness,
* Odd position in water, head or tail down, difficulty maintaining buoyancy,
* Fish gasping at the surface,
* Fish rubbing or scraping against objects.

Abiotic Diseases: Most of the mortalities in tilapia farming are not caused by pathogens, but rather by abiotic causes primarily related to water quality or toxicity. Poor water quality conditions can induce opportunistic infections that can easily occur in unhealthy or stressed fish.

Biotic Diseases: In general, tanks with a recirculating filter system are less affected than pond or cage aquaculture farming by pathogens. In most cases, pathogens are actually already present in the system, but disease does not occur because the fishes' immune system is resisting infection and the environment is unfavorable for the pathogen to thrive.

Healthy management, stress avoidance, and quality control of water are thus necessary to minimize any disease incidence. Whenever disease occurs, it is important to isolate or eliminate the infected fish from the rest of the stock and implement strategies to prevent any transmission risk to the rest of the stock. If any cure is put into action, it is fundamental that the fish be treated in a quarantine tank, and that any products used are not introduced into the fish tank or pond. More details on this subject are available from regulatory agencies, such as the U.S. Fish and Wildlife Service, online research, and most local fishery organizations.

Beware of Fish Disease Treatments

Antibiotics and copper treatments should ideally only be used in a quarantine situation (a separate quarantine tank), and the water from the quarantine tank should never be added to the fish tank or pond. All recommended treatments need to be carefully scrutinized to ensure that they are pertinent to the symptoms you see in your fish, and are appropriate for the age of the fish.

Steps to Treating Sick Fish

If a significant percentage of fish are showing signs of disease, it is likely that the environmental conditions are causing stress. In these cases, check levels of ammonia, nitrite, nitrate, pH, and temperature in order to respond accordingly. If only a few fish are affected, it is important to remove the infected fish immediately in order to prevent the spread of the any disease to the other fish. Once removed, inspect the fish carefully and attempt to determine the specific disease and cause.

However, it may be necessary to have a professional diagnosis carried out by a veterinarian, extension agent, or other aquaculture expert. Knowing the specific disease helps to determine the treatment options. Place the affected fish in a separate tank, (referred to as a "quarantine" or "hospital tank"), for further observation. Kill and dispose of the fish if appropriate.

Commercial drugs can be expensive and/or difficult to procure. If treatment is absolutely necessary, it should be done in a hospital tank only; antibacterial chemicals should not be added to the fish tank. One effective treatment option against some of the most common bacterial and parasite infections is a salt bath.

Salt Bath Treatment: Fish affected with some ectoparasites, molds, and bacterial gill contamination can benefit from salt bath treatment. Infected fish can be removed from the main fish tank or pond and placed into a salt bath. This salt bath is toxic to the pathogens, but non-fatal to the fish. The salt concentration for the bath should be 1 kg of salt per 100 liters of water. Affected fish should be placed in this salty solution for 20–30 minutes, and then moved to a second isolation tank containing 1–2 g of salt per liter of water for another 5–7 days.

With bad white-spot infections, all fish may need to be removed from the main tank and treated this way for at least a week. During this time, any emerging parasites in the tank will fail to find a host and eventually die. Heating of the water can also shorten the parasite life cycle and make the salt treatment more effective. Do not implement a salt bath treatment within the fish tank, or use any salt bath water when moving the fish back into the main tank as the salt concentrations would negatively affect the system infrastructure and other fish within the system.

PART II : FISH

CHAPTER 5

Fish Feed

Fish Feed and Nutrition

Fish require the correct balance of proteins, carbohydrates, fats, vitamins, and minerals to grow and be healthy. This type of feed is considered a whole feed. Commercial fish feed pellets are readily available from numerous sources from online suppliers to local feed stores.

Protein is the most important component for building fish mass. In their grow-out stage, omnivorous fish, such as tilapia, need 25–35 percent protein in their diet, while carnivorous fish need up to 45 percent in order to grow at optimal levels. In general, younger fish (fry and fingerlings) require a diet richer in protein than during their grow-out stage. Proteins are the basis of structure and enzymes in all living organisms. Proteins consist of amino acids, some of which are synthesized by the fishes' bodies, but others have to be obtained from the food. These are called essential amino acids. Of the ten essential amino acids, methionine and lysine are often limiting factors, and these need to be supplemented in some vegetable-based feeds.

Lipids are fats, which are high-energy molecules necessary to a fish's diet. Fish oil is a common component of fish feeds. Fish oil is high in two special types of fats, omega-3 and omega-6, that have health benefits for humans. The amount of these healthy lipids in farmed fish depends on the feed used.

Carbohydrates consist of starches and sugars. This component of the feed is an inexpensive ingredient that increases the energy value of the feed. The starch and sugars also help bind the feed together to make a pellet.

Vitamins and minerals are necessary for fish health and growth. Vitamins are organic molecules, synthesized by plants or through manufacturing, that are important for development and immune system function. Minerals are inorganic elements. These minerals are necessary for the fish to synthesis their own body components (bone), vitamins, and cellular structures. Some minerals are also involved in osmotic regulation.

Pelletized Fish Feed

There are a number of different sizes of fish feed pellets. Obviously, the recommended size of pellet depends on the size of the fish. Fry and fingerlings have small mouths and cannot ingest large pellets, while large fish waste energy if the pellets are too small. If possible, the feed should be purchased for each stage of the lifecycle of the fish. Alternatively, large pellets can be crushed with a mortar and pestle to create powder for fry and crumbles for fingerlings. Some operators use the same medium-sized pellets (2–4 mm) so that their fish continue to eat the same-sized pellet from the fingerling stage right up to maturity.

Fish feed pellets are also designed to either float on the surface or sink to the bottom of the tank, depending on the feeding habits of the fish. It is important to know the eating behavior of your specific fish and supply the correct type of pellet. Floating pellets are advantageous because it is easier to identify how much the fish are eating. It is often possible to train fish to feed according to the food pellets available; however, some fish will not change their feeding culture.

Feed should be stored in dark, dry, cool, and secure conditions. Fish feed will attract rodents and other vermin, if not securely contained. Warm wet fish feed can rot, being decomposed by bacteria and fungi. These micro-organisms can release toxins that are dangerous to fish; spoiled feed should never be fed to fish. Fish feed should not be stored for too long, should be purchased fresh, and used immediately to conserve the nutritional qualities, wherever possible.

Avoid overfeeding your fish. Uneaten food waste should never be left in the tank. Feed waste from over-feeding is consumed by heterotrophic bacteria, which devours substantial amounts of oxygen. In addition, decomposing food can increase the amount of ammonia and nitrite to toxic levels in a relatively short period. Finally, the uneaten pellets can clog the mechanical filters, leading to decreased water flow and anoxic areas. In general, fish eat all they need to eat in a 30-minute period. If uneaten food is found, lower the amount of feed given at the next feeding.

Fish Feed

The following provides some general information about fish feed:

- Live foods are a good source of supplements and provide variety.
- Fish typically eat all their food within 3 minutes. If feed remains after 10 minutes, it is a good indication that they are being overfed. Overfeeding will lead to food decomposition toxicity problems. Feeding a variety of different feedbrands helps ensure proper nutrition.
- Most commercial fish feeds contain exact protein, carbohydrate, and other vitamin requirements for specific fish.
- Plant based proteins can include soy meal, corn meal, wheat meal, etc.
- Most commercial feeds are between 10 to 35 percent protein.
- Alternative feeds should be considered like duckweed, insects, worms, or black soldier fly larvae.
- Avoid fishmeal based feeds as this source is not derived from environmental friendly sustainable practices. Fishmeal feed can also come from animals raised on growth hormones, antibiotics, pesticides, herbicides, and fungicide laced feed, which then enters your food supply.
- Most fish feed is GMO based, which can entail the presence of heavy metals. In turn, this is passed on up the food chain to you and your family in higher concentrations. Heavy metals wreak havoc on our health and well-being. Therefore, try to use only fish feed that is made entirely from non-GMO, USDA certified organic ingredients, is free of terrestrial animal parts, and contains no fish meal or soy. There are many suppliers of organic fish feed. An Internet search for organic fish feed will provide numerous options.
- Many online vendors will provide you with the nearest retail outlet that sell their feed, saving you the shipping cost. Depending upon your location, though, having it shipped to you may be the only option. Teaming up with other aquaculture operators in your area could enable you to get better bargains via buying in bulk and sharing the shipping cost.

Feeding Fish — Parameters and Other Considerations

In some operations, feed costs can add up to half or more of the total annual operating costs. For this reason, it is import ant to use the proper feed and to take measures to be sure that the conversion of feed to fish flesh is efficient. The Feed Conversion Ratio (FCR) describes how efficiently an animal turns its food into growth. It answers the question of how many units of feed are required to grow one unit of animal. FCRs exist for every animal and offer a convenient way to measure the efficiency and costs of raising any given animal.

In working with fish and fish feed it is easier to use metric units, and such is the industry standard, even in the United States. Fish, in general, have one of the best FCRs of all livestock. The FCR is expressed as the number of pounds of feed required to produce one pound of flesh. Depending on the species being grown, it may take 1.75 to produce 1 pound of fish. In good conditions, tilapias have an FCR of 1.4–1.8, meaning that to grow a 2.2 pound (1.0 kg) tilapia, 3-4 pounds (1.4–1.8 kg) of food is required.

Tracking FCR is not essential in small-scale aquaponics, but it can be useful to do in some circumstances. When changing feeds, it is worth considering how well the fish grow in regard to any cost differences between the feeds. Moreover, when considering starting a small commercial system, it is necessary to calculate the FCR as part of the business

FIGURE 20. Weighing a sample of fish using a weighing scale

Simple Steps for Weighing Fish

1. Fill a small bucket with 2.5 gallons (10 liters) of water from the aquaponic system.
2. Place the bucket on a weighing scale and record the weight (tare).
3. Scoop 5 average size fish with a landing-net, drain the landing-net of excess of water for a few seconds then place the fish into the bucket.
4. Weigh again and record the gross weight.
5. Calculate the total weight of the fish by subtracting the tare from the gross weight.
6. Divide this figure by 5 to retrieve an average weight for each fish.

Repeat steps 1–6 as appropriate. Try to measure 10–20 percent of the fish (preferably no duplicates) for an accurate average.

plan and/or financial analysis. Even if not concerned about the FCR, it is good practice to periodically weigh a sample of the fish to make sure they are growing well and to understand the balance of the system.

This also provides a more accurate growth rate expectation for harvest timing and production. As with all fish handling, weighing is easier in darkness to avoid stressing the fish. Following is a list of simple steps for weighing fish. Weighing fish of the same age, growing in the same tank, is more preferable than heterogeneous cohorts of fish as the measurement should provide more reliable averages.

Periodic weight measurements will give the average growth rate of the fish, which will be obtained by subtracting the average fish weight, calculated above, over two periods. The FCR is obtained by dividing the total feed consumed by the fish by the total growth during a given period, with both values expressed in the same weight unit (i.e., ounces and pounds, or gram or kilogram).

Total Feed / Total Growth = FCR

The total feed can be obtained by summing all the recorded amount of feed consumed each day. The total growth can be calculated by multiplying the average growth rate by the number of the fish stocked in the tank.

At the grow-out stage, the feeding rate for most cultured fish is 1.5–2 percent of their body weight per day. On aver- age, a 3.5-ounce (100 gram) fish eats 0.04 to 0.07 ounces (1–2 grams) of pelletized fish feed per day.

For example, if you have 75 lbs. (34kg) of fish in your tank, multiply 75 lbs. x 1.5% (0.015) = 1.125 lbs. of fish feed daily (34kg x 1.5% (0.015) = 0.5kg).

If the fish still appear to be consuming all of the feed, then increase the amount to 1.75 percent and then to 2 percent (a multiplier of 0.0175 and 0.02, respectively). As mentioned in the previously, in working with fish and fish feed, it is easier to use metric units; such is the industry standard, even in the United States.

Monitor the feeding rate and FCR to determine growth rates and fish appetite helps maintain overall system balance. However, don't just rely on the calculations. Observe your fish eating to help determine the proper amount of feed needed above or below the recommended amount. If they rapidly devour all of the feed, they most likely need more. If the fish are showing little interest and much of the feed is being wasted on the bottom of the tank or pond, then they are probably being overfed.

As a friendly reminder to an important and often overlooked issue mentioned earlier in this chapter, in selecting aproper feed, it is important to match the feed size to the size of the fish being fed. Smaller particle or pellet size is required for smaller fish, while larger pellets will be more efficiently used by larger fish.

The feed must be suitable to meet the nutritional needs of the species being cultured. A commercially manufactured feed is available for most species.

Feeding rates will vary by fish size, stocking density, and operation system, but a few common factors contribute to determining these rates:
- Water temperature.
- Water quality.
- Size of the feed particles.
- Palatability of the feed to the fish.
- Frequency of feeding.
- Technique for feed delivery.
- Type of feed used (i.e., floating, sinking, etc.).

Feed can be delivered in a variety of ways, including feeding by hand, using automated stationary feeders, or by allowing the fish to feed themselves using "demand" feeders. The choice of feed delivery system used will be dictated by logistical needs, resources available, amount of feed to be delivered, number of fish to be fed daily, and the size, scope, and type of operation.

TABLE 6. **A Good Trout (Tilapia) Feed Compostion**

GUARANTEED ANALYSIS	
Crude Protein, min.	40%
Crude at, min	12%
Crude Fiber, max	3%
Ash, max	12%

INGREDIENTS

Fish Meal, Soybean Meal, Wheat Flour, Stabilized Fish Oil, Wheat Midds, Poultry By-Product Meal, Blood Meal, Hydrolized Feather Meal, Corn Gluten Meal, Poultry Oil, Vitamin A Acetate, D-Activated Animal Sterol (D_3), Vitamin B12 Supplement, Riboflavin Supplement, Niacin, Folic Acid, Menadione Sodium Bisulphite Complex, Calcium Pantothenate, Pyridoxine Hydrochloride, Thiamine, Biotin, DL Alphatocopherol (E), L-ascorbyl-2-polyphosphate (C), Betaine, Zinc Sulfate, Copper Sulfate, Ferrous Sulfate, Manganese Sulfate, Ethylenediamine Dihydriodide, Ethoxyquin (Anti-Oxidant).

Keep fry (newborns) separated from bigger fish so they are not eaten. Feed fry a diet of micro worms (nematode), brine shrimp, or soaked oatmeal (soft things). Feed fingerlings (between newborn and mid-grown) small fish flakes.

Extruded trout feed is a proven favorite of the North American Trout industry. Formulated to allow controlled growth and deliver excellent feed conversions, extruded trout feed can lead to more efficient production. It can be fed to numerous species including Trout, Perch, Bass, Sturgeon, Catfish, and Tilapia. A typical good feed composition for tilapia is shown in Table 6.

Practical Aspects of Feeding Fish

Hand-feeding Techniques

Hand and mechanized feedings are the two widely practiced techniques. Of these, hand feeding is the recommended one. Calibrated spoons and hand shovels should be used in order to ensure exact and uniform portions of feed.

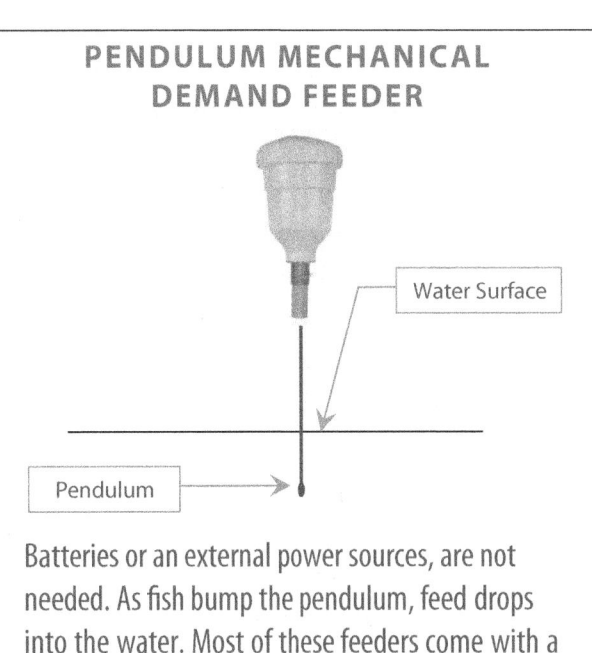

PENDULUM MECHANICAL DEMAND FEEDER

Batteries or an external power sources, are not needed. As fish bump the pendulum, feed drops into the water. Most of these feeders come with a unique twist-lock lid, which is wind- and varmint-proof. Furthermore, the better feeders are made with UV-resistant polyethylene. There are four common sizes, which accommodate #4 crumble to ¼-inch pellets.

Typical Cost (year 2020, USD)

Demand Feeder 22 lbs (10 Kg)	$392.25
Demand Feeder 44 lbs (20 Kg)	$474.63
Demand Feeder 88 lbs (40 Kg)	$518.25
Demand Feeder 132 lbs (60 Kg)	$743.36

FIGURE 21. *Pendulum Mechanical Demand Feeder*

Loss of appetite among fish is one of the most obvious symptoms of many different problems. It indicates, among others concerns, insufficient oxygen content of water or a developing disease in fish. Therefore, regular daily feeding is an excellent opportunity to observe fish, detect problems, and diagnose diseases.

Demand and Automatic Feeders

Demand feeders are those that release feed according to the appetite of fish. Because some fish species are very greedy, these feeders may allow unnecessary overfeeding of fish unless the portions are controlled.

The advantage of mechanized and automatic feeders is that they save on labor and can allow the operator to have a more flexible schedule. The most typical mechanized and automatic feeders are the demand bar feeder, used for fish size 50 g, and the clock-driven feeding belt.

Signs of Feeding Problems

Obvious signs of feeding problems are the increasing differences in individual sizes, growing aggressiveness, and cannibalism. Lack of sufficient feed manifests itself in bitten/damaged fish and dead fish.

Economics of Feeding Fish

Economics of production is related primarily to efficiency of the system and market price. Two of the most predictable and significant variable costs are fish and feed. Feed cost to produce a one-pound fish is related to conversion rate and cost of the feed.

Table 7 on the next page shows cost of feed per pound of fish meat produced. Tilapia typically require from 1.2 to 1.7 lb. of feed for one pound of gain for food size animals. Please note that the costs below are based upon buying in bulk and are geographic specific. Prices will vary when buying in smaller quantities and your location.

Reducing Feed Cost

To offset cost, many tilapia farm operators report having success growing earth and blood worms separately to feed to their fish. Some operators maintain a nearby compost bin for worms or raise solder grubs. Others grow crickets, roaches, or other insects as feed. Some operators even grow maggots.

If the fish tank is located outdoors, a low voltage submersible LED light can be integrated into the fish tank to attract bugs at night. If implementing this

TABLE 7. **Estimating Feed Cost/lb Gain**

Cost/Ton	Cost/50 lb bag	Cost/lb 0.9	Feed Conversion (lb feed required for lb gain)										
			1	1.1	1.2	1.3	1.4	1.5	1.6	1.7	1.8	1.9	2
$400	$10	$0.20	$0.20	$0.22	$0.24	$0.26	$0.28	$0.30	$0.32	$0.34	$0.36	$0.38	$0.40
$440	$11	$0.22	$0.22	$0.24	$0.26	$0.29	$0.31	$0.33	$0.35	$0.37	$0.40	$0.42	$0.44
$480	$12	$0.24	$0.24	$0.26	$0.29	$0.31	$0.34	$0.36	$0.38	$0.41	$0.43	$0.46	$0.48
$520	$13	$0.26	$0.26	$0.29	$0.31	$0.34	$0.36	$0.39	$0.42	$0.44	$0.47	$0.49	$0.52
$560	$14	$0.28	$0.28	$0.31	$0.34	$0.36	$0.39	$0.42	$0.45	$0.48	$0.50	$0.53	$0.56
$600	$15	$0.30	$0.30	$0.33	$0.36	$0.39	$0.42	$0.45	$0.48	$0.51	$0.54	$0.57	$0.60
$640	$16	$0.32	$0.32	$0.35	$0.38	$0.42	$0.45	$0.48	$0.51	$0.54	$0.58	$0.61	$0.64
$680	$17	$0.34	$0.34	$0.37	$0.41	$0.44	$0.48	$0.51	$0.54	$0.58	$0.61	$0.65	$0.68
$720	$18	$0.36	$0.36	$0.40	$0.43	$0.47	$0.50	$0.54	$0.58	$0.61	$0.65	$0.68	$0.72
$760	$19	$0.38	$0.38	$0.42	$0.46	$0.49	$0.53	$0.57	$0.61	$0.65	$0.68	$0.72	$0.76
$800	$20	$0.40	$0.40	$0.44	$0.48	$0.52	$0.56	$0.60	$0.64	$0.68	$0.72	$0.76	$0.80
$840	$21	$0.42	$0.42	$0.46	$0.50	$0.55	$0.56	$0.63	$0.67	$0.71	$0.76	$0.80	$0.84
$880	$22	$0.44	$0.44	$0.48	$0.53	$0.57	$0.62	$0.66	$0.70	$0.75	$0.79	$0.84	$0.88
$920	$23	$0.46	$0.46	$0.51	$0.55	$0.60	$0.64	$0.69	$0.74	$0.78	$0.83	$0.87	$0.92

method, be sure to shield the light so that the fish are not exposed to any direct lighting.

Duckweed is a popular way to feed Tilapia. Duckweed can be raised by those of the do-it-yourself crowd or purchased from a multitude of sources. Duckweed is addressed comprehensively later in this chapter.

It is perfectly acceptable to feed your fish store bought food or table scraps so long as it satisfies to their dietary needs. This approach will help lower your feed cost.

Some fish only eat a plant-based diet, while some species only eat smaller animals (fish, worms, insects, cut-up meat scraps, etc.), and some fish eat both. In nature, tilapia eat both, but most tilapia farmers have their fish on a herbaceous diet.

Real World Commercial Aquaculture Data Farming Striped Bass

The following data is based on a real-world hybrid striped bass commercial aquaculture businesses operating in the United States using a recirculating tank system. Labor, oxygen, and biocaronate line items used in commercial aquaculture operations have been omitted to better reflect a commercial aquaponic operation. Hybrid striped bass are generally harvested at a weight of 1.5 to 2.5 pounds when they are 18 to 24 months old. In controlled growing environments water temperature and quality can be controlled, not ponds where conditions are subject to change, growth is faster (harvest at 18 months). Reducing the above-mentioned commercial aquaculture operation down to the typical size of a small backyard aquaponic operation (raising bass) in a 250-gallon fish tank, the economics would more closely resemble the following for one harvest cycle:

$200	Total Operational Cost.
$8	Cost per lb. (Cost is higher per lb. in smaller systems).
26.5	Pounds of ending biomass.
$13.54	Average retail market value for striped bass.
$5.54	Net profit per lb.
$146.81	Total Net Profit (250-gallon fish tank).

In some areas of the United States, such as parts of northern California for instance, it is illegal to have Tilapia, if they are located in indoor tanks; thus, shipping of Tilapia to the areas is prohibited. Therefore, it is important to check with your regional US Fish and Game office, or the governing regulatory agency in your region, to ensure that you are in compliance with all applicable laws in your area. Desiring to raise a species that is not common in your area or country will likely be very expensive to obtain.

Tilapias are among the most popular fish for those that are into aquaculture as a hobby or as an extra source of low-cost family food. Tilapias are much more forgiving of water temperature and pH changes (mistakes). They also have a much more diversified diet and reach harvest size a lot sooner than other food fish.

NOTE: Most backyard aquaculture systems have fish tanks ranging in size from about 250 to 1,000 gallons.

TABLE 8. **Fish Tank Efficiency**

Water volumne, galloons	2,500
Size stocked (grams)	1
Size harvested (grams)	567
Size harvested (lbs)	1.25
Survival rate	85%
Feed cost, per pound	$0.52

Commercial Fish Feed Problem and a Possible Solution

About half of the world's seafood now comes from fish farms. From the environmental perspective, that is creating a major problem: millions of tons of wild fish like anchovies, sardines and mackerel are being caught in the ocean to feed farm-raised fish like salmon. Most anchovies and sardines don't end up on pizzas. Instead, they go to processing plants where they are turned into pellets to feed farmed fish. We are depleting the world's oceans to make a cheap protein.

Up to 90 percent of tiny, harvested forage fish from the oceans go into pet food, poultry feed and fishmeal, never destined for human consumption. Small, filter feeding fish were traditionally used because they were inexpensive. But as wild fish stocks diminish, the cost of these forage fish increases. Meanwhile, the price for plants like corn and soy has decreased.

Scientists and entrepreneurs are beginning to find ways to create vegetarian diets for species like trout, which may lessen the strain on over-fished oceans. To avoid using wild fish in farmed fish diets, the United States Department of Agriculture has spent the past ten years researching alternative diets that include plants, animal processing products, insects and single-cell organisms like yeast, bacteria, and algae. "We have been hit over the head with the notion that farming carnivorous fish means that you have to catch fish in the ocean for its diet, but that's wrong," says Michael Rust, science coordinator of the NOAA Fisheries' Office of Aquaculture.

Recent advances in aquaculture research have shown that farmed carnivorous fish do not require any fishmeal or fish oil in their feeds. The USDA has proven that many species of fish can get enough nutrients from these alternative sources without eating other fish. If widely adopted by the aquaculture industry, this plant-based diet could significantly reduce the amount of wild fish that are harvested and turned into fish meal pellets.

The research comes at a pivotal time when a growing population will mean an increased demand for seafood. Americans as a whole consumed 5.9 billion pounds of seafood in 2018 and to meet that demand, 91 percent of it was imported. On average, each American consumed 16.1 pounds of seafood in 2018.

Without increased aquaculture production, the world will face a seafood shortage of 50 to 80 million tons by 2030, according to the United Nations Food and Agriculture Organization. But that production needs to be sustainable by decreasing the use of wild fish according to the USDA.

Fish, like people, don't need specific foods but rather specific nutrients in order to stay healthy. In fact, all animals essentially need the same forty nutrients—a combination of amino acids, fatty acids, vitamins and minerals. With this in mind a pair of entrepreneurs started a company called 'TwoxSea'. They have found that aquaculture facilities were concerned about the environmental impact of traditional fish feed, future availability, and cost of such; and as result were very open to replacing fishmeal with alternative feeds like corn, soy and algae. They now have several large commercial customers feeding their fish nothing but TwoxSea's nutritionally complete vegetarian feed. By all accounts, there is now hope that we can have a growing aquaculture industry that is sustainable, and we won't have to rely on the ocean to get our fish.

Alternative Fish Feed

Fish feed is one of the most important and expensive inputs of tilapia farming. It can be purchased or self-made. Purchasing a quality manufactured whole food fish feed is certainly the easiest way to go and ensures the nutritional needs of the fish are being met. Even so, below is an example of supplemental fish feed that can be easily produced domestically, which can help save money or used temporarily if manufactured feeds are not available.

Duckweed

Duckweed is the second smallest flowering plant in the world (watermeal being the smallest). It floats and grows directly on top of water. Not all fish take to duckweed immediately, but most herbivorous fish adjust quickly. Duckweed grows in calm waters. It produces oxygen for the water when in sunlight and consumes oxygen from the water when shaded and at night. Duckweed is a fast-growing floating water plant that is rich in protein and can serve as a food source for tilapia. It is a useful addition to a Tilapia farm system.

Duckweed can double its mass every 1–2 days in optimum conditions, which means that one-half of the duckweed can be harvested every day. Duckweed should be grown in a separate tank from the fish because otherwise the fish would consume the whole stock. Aeration is not necessary, and water should flow at a slow rate through the container in which the duckweed is grown. Duckweed can be grown in sun-exposed or half-shaded places. Surplus duckweed can be stored and frozen in bags for later use. Duckweed is also a useful feed for poultry.

Duckweed consumes more nutrients than most plants. Duckweed can consist of up to a 45 percent protein, which even surpasses the protein concentration of soybeans. It also has all the essential amino acids. Duckweed can be obtained from a variety of sources online simply by performing an Internet search.

Azolla (Water Fern)

Azolla is a genus of fern that grows while floating on the surface of the water, much in the manner as duckweed. The major difference is that Azolla is able to fix atmospheric nitrogen; essentially creating protein from the air. This occurs because Azolla has a symbiotic relationship with a species of bacteria, Anabaena azollae, which is contained within the leaves.

As well as providing a free source of protein, Azolla is an attractive feed source because of its exceptionally high growth rate. Like duckweed, Azolla should be grown in a separate tank with slow water flow. Its growth is often limited by phosphorus, so if Azolla is to be grown intensively an additional source of phosphorous is needed such as compost tea.

FIGURE 22. Growing duckweed for fish feed for mostly herbivorous fish, such as Tilapia, Goldfish, and Koi.

FIGURE 23. Azolla spp. growing in a container as fish feed supplement

Insects

Insects are considered undesirable pests in many cultures. However, they have an enormous potential in supporting traditional food chains with more sustainable solutions. In many countries insects are already part of people's diets and sold at the markets. In addition, they have been used as animal feed for centuries.

Insects are a healthy nutrient source because they are rich in protein and polyunsaturated fatty acids and full of essential minerals. Their crude protein content ranges between 13 and 77 percent (on average 40 percent) and varies according to the species, the growth stage, and the rearing diet. Insects are also rich in essential amino acids, which are a limiting factor in many feed ingredients. Edible insects are a good source of lipids, as their quantity of fat can range between 9 and 67 percent. In many species, the content of essential polyunsaturated fatty acids is also high. These characteristics together make insects a healthy and ideal option for both human food, and feed for animals or fish.

Given their enormous number and varieties, the choice of the insect to be reared can be tailored to their local availability, climatic conditions/seasonality and type of feed available. The source of food for insects can include staple husks, vegetable leaves, vegetable wastes, manure, and even wood or cellulose-rich organic materials, which are suitable for termites. Insects also make a great contribution to waste biodegradation, as they break down organic matter until it is consumed by fungi and bacteria and mineralized into plant nutrients.

The culturing of insects is not as challenging as other animals. Furthermore, minimal space is needed to raise a large insect population. Sometimes insects are referred to as "micro-livestock". The small space requirement means that insect farms can be created with extremely limited areas and with only a minor investment cost. In addition, insects are cold-blooded creatures, this means that their feed conversion efficiency into meat is much higher than terrestrial animals and similar to fish. Lots of possible options and additional knowledge on insect farming as feed is available on the internet. Among the many species available, an interesting one to be used as fish feed is the black soldier fly.

Black Soldier Fly

The larvae of black soldier flies, Hermetia illucens, are extremely high in protein and thus, a valuable source for various livestock types, including fish. The lifecycle of this insect makes it a convenient and attractive addition to an integrated homestead farming system in favorable climate conditions. The larvae feed on manure, dead animals, and food waste. When culturing black soldier flies, these types of waste are placed in a compost unit that has adequate drainage and airflow. As the larvae reach maturity, they crawl away from their feed source through a ramp installed in the compost unit that leads to a collection bucket.

Essentially, the larvae devour wastes, accumulate protein, and then harvest themselves. Two-thirds of the larvae can be processed into feed while the remaining one-third should be allowed to develop into adult flies in a separate area. The adult flies are not a vector of disease; adult flies do not have mouthparts, do not eat, and are not attracted to any human activities. Adult flies simply mate and then return to the compost unit to lay eggs, dying after a week. Black soldier flies have been shown to prevent houseflies and blowflies in livestock facilities and can actually decrease the pathogen load in the compost. Even so, before feeding the larvae to the fish, the larvae should be processed for safety. Baking in an oven (170 °C for 1 hour) destroys any pathogens, and the resulting dried larvae can be ground and processed into a feed.

Moringa or Kalamungay

Moringa oleifera is a species of tropical tree that is remarkably high in nutrients, including proteins and vitamins. It is a fast-growing, deciduous tree that can reach a height of 32–40 ft (10–12 m). Classified by some as a super food and currently being used to combat malnutrition, it is a valuable addition to homemade fish feeds because of these essential nutrients. All parts of the tree are choice edibles suitable for human consumption, but for aquaculture it is typically the leaves that are used. In fact, there has been success in several small-scale aquaculture projects in Africa using leaves of this tree as the only source of feed for tilapia.

These trees are drought-resistant, and easily propagated through cuttings or seeds. However, they are intolerant of frost or freezing and not appropriate for cold areas. For leaf production, all of the branches are harvested down to the main trunk four times per year in a process called pollarding.

Making Homemade Fish Feed

As mentioned previously, fish feed is one of the most expensive inputs in aquaculture. Feed is also one of the most important components of the system because it is what sustains the fish. In addition, toxic components in used some in feed can work up the food chain in higher concentrations. Therefore, it is necessary that operators understand its composition. Also, if commercial pelleted feed is not available, it is important to understand how to make it.

FIGURE 24. Black soldier fly (Hermetia illucens) *adult (a) and larvae (b)*

Specific Nutritional Requirements for Tilapia

The low trophic level and the omnivorous food habits of tilapia make them a relatively inexpensive fish to feed, unlike other finfish, such as salmon, which rely on high-protein and lipid diets based on more expensive protein sources like fish meal. In addition, tilapias are similar to channel catfish (Ictalurus punctatus), in that they can tolerate higher dietary fiber and carbohydrate concentrations than most other cultured fish. To ensure high yield and fast growth at least cost, a well-balanced prepared feed is essential to successful tilapia culture. Slight variations exist among tilapia species, but nutrient requirements are primarily affected by the size of the fish.

Composition of Fish Feed

Fish feed consists of all the nutrients that are required for growth, energy, and reproduction. Dietary requirements are identified for proteins, amino acids, carbohydrates, lipids, energy, minerals, and vitamins. A brief summary of major feed components is listed as follows.

Protein

Dietary protein plays a fundamental role for the growth and metabolism of animals. A combination of more than 100 amino acids joined by peptides forms a protein; they are the building blocks of protein. Only some amino acids can be synthesized by animals while others cannot, so these must be supplied in the diet. Non-essential amino acids can be synthesized internally. However, this does not mean they are unimportant. It is just that the body is capable of producing a sufficient amount to meet the demands for growth and tissue repair. Essential amino acids cannot be synthesized by the body and must be acquired through outside sources.

For aquatic animals, there are ten essential amino acids (EAAs): arginine, histidine, isoleucine, leucine, lysine, methionine, phenylalanine, threonine, tryptophan, and valine. Therefore, feed formulation must find an optimal balance of EAAs to meet the specific requirements of each fish species. Non-compliance with this requirement would prevent fish from synthesizing their own proteins, and also waste the amino acids that are present. The ideal feed formulation should thus take into account the EAA levels of each ingredient and match the quantities required by fish.

The profile of dietary protein is important when formulating diets for tilapia. Recommended protein intake of fish depends on the species and age. For tilapia and herbivorous fish, the optimal ranges are 28–35 percent; carnivorous species require 38–45 percent. Juvenile fish require higher-protein diets than adults due to their intense body growth.

TABLE 9. **Protein requirements (dry basis) of tlapia of various species and sizes.**

Species	Size (g)	Requirement (% of diet)	Reference
O. niloticus	0.5-68.3	40	Al Hafid (1999)
	45.0-76.3	40	
	0.84-22.8	40	Siddiqui et al. (1988)
	40.4-108.0	30	
O. niloticus x O. aurevs	21.3-81.5	28	Twibelll and Brown (1998)
O. aurevs	2.5-16.6	56	Winfree and Stickney (1981)
S. mossambicus	1.83-8.5	40	Jauncey (1981)

TABLE 10. **Amino acid requirements (dry basis) of tilapia.**

Amino acid	% of dietary protein	
	O. niloticus"	O. mossambicus>
Arginine	4.20	2.82
Histidine	1.72	1.05
Isoleucine	3.11	2.01
Leucine	3.39	3.40
Lysine	5.12	3.78
Methionine	2.68	0.99
Phenylalanined	3.75	2.50
Threonine	3.75	2.93
Tryptophan	1.00	0.43
Valine	2.80	2.20

Dietary proteins are used continuously by fish for maintenance, growth, and reproduction functions. When fed in excess, protein may be used as energy; however, the latter function is not desirable because of the expensive cost of proteins. The protein requirement of tilapia decreases with age and size, with higher dietary CP concentrations required for **tilapia fry (30–56%)** and **juvenile (30–40%)** tilapia but lower protein levels **(28–30%)** for larger tilapia.

Besides any optimal amino acid content in the feed, it is worth stating the importance of an optimal dietary balance between proteins and energy (supplied by carbohydrates and lipids) to obtain the best growth performance and reduce costs and wastes from using proteins for energy. Although proteins can be used as a source of energy, they are much more expensive than carbohydrates and lipids, which are preferred.

Carbohydrates

Carbohydrates provide a relatively inexpensive source of energy compared to protein, and their inclusion can improve the quality of pelleted feeds. A matter of fact, tilapia can effectively utilize carbohydrate levels up to 30 to 40% in the diet, which is considerably more than most cultured fish.

Carbohydrates are mainly composed of simple sugars and starch, while other complex structures such as cellulose and hemicellulose are not digestible by fish. In general, the maximum tolerated amount of carbohydrates should be included in the diet in order to lower the feed costs. Omnivorous fish (tilapia) can easily digest quantities up to 40 percent, but the percentage falls to about 25 percent in carnivorous and cold-water fish. Carbohydrates are also used as a binding agent to ensure the feed pellet keeps its structure in water. In general, one of the most used products in extruded or pelleted feed is starch (from potato, corn, cassava, or gluten wheat), which undergoes a gelatinization process at 140-185°F (60–85 °C) that prevents pellets from easily dissolving in water.

Lipids

Lipids provide energy and essential fatty acids (EFAs), indispensable for the growth and other biological functions of fish. Fats also play the important role in absorbing fat-soluble vitamins and securing the production of hormones. Fish, as with other animals, cannot synthesize EFAs, which have to be supplied with the diet according to the species' needs. Deficiency in the supplement of fatty acids results in reduced growth and limited reproductive efficiency.

In general, freshwater fish require a combination of both omega-3 and omega-6 fatty acids, whereas marine fish need mainly omega-3. Tilapias mostly require omega-6 in order to secure optimal growth and high feed conversion efficiency.

Dietary lipids provide a major source of energy, facilitate the absorption of fat-soluble vitamins, play an important role in membrane structure and function, serve as precursors for steroid hormones and prostaglandins, and serve as metabolizable sources of essential fatty acids. For tilapia up to 2.5 g, the optimum dietary lipid concentration is 5.2%, decreasing to 4.4% for fish up to 7.5 grams. **To maximize protein utilization, dietary fat concentration should be between 8 and 12% for tilapia up to 25 grams, and 6 to 8% for larger fish.** As with most fish, tilapia appear to have a requirement for n-6 (linoleic) fatty acids, and to a lesser extent, a requirement for n-3 (linolenic) fatty acids. Studies have shown that dietary lipids should supply at least 1% of n-6 fatty acids. When dietary lipids contain considerable amount of polyunsaturated fatty acids, attention should be given to prevent oxidation of the dietary lipids. The resulting products of lipid oxidation are toxic, reduce the availability of other nutrients, and impact the quality of fish-flesh products.

Energy

Energy is mainly obtained by the oxidation of carbohydrates, lipids and, to a certain extent, proteins. The energy requirements of fish are much lower

TABLE 11. **Common feed ingredient sources of the most important nutrient components**

NUTRIENT COMPONENTS	FEED INGREDIENT SOURCES
Protein	*Plant-based sources*: algae, yeast, soybean meal, cottonseed meal, peanuts, sunflower, rapeseed/canola, other oil-seed cakes *Animal-based sources*: fishery by-products (fishmeal or offal), poultry by-products (poultry meal or offal), meat meal, meat and bone meal, blood meal
Carbohydrates	Wheat flour, wheat bran, corn flour, corn bran, rice bran, potato starch, cassava root meal
Lipids	Fish oil, vegetable oil (soybean, canola, sunflower), processed animal fat
Vitamins	Vitamin premix, yeast, legumes, liver, milk, bran, wheat germ, fish and vegetable oil
Minerals	Mineral premix, crushed bone

than warm-blooded animals owing to the reduced needs to heat the body and to perform metabolic activities. However, each species requires an optimum amount of protein and energy to secure best growth conditions and to prevent animals from using expensive protein for energy. It is thus important that feed ingredients be carefully selected to meet the desired level of digestible energy (DE) required by each aquatic species.

Vitamins and Minerals

Vitamins are organic compounds necessary to sustain growth and to perform all the physiological processes needed to support life. Vitamins must be supplied with the diet because animals do not produce them. Vitamin deficiencies are most likely to occur in intensively cultured cages and tank systems where animals cannot rely on natural food. Degenerative syndromes are often ascribed to an insufficient supply of these vitamins and minerals.

Minerals are important elements in animal life. They support skeletal growth and are also involved in osmotic balance, energy transport, neural, and endocrine system functioning. They are the core part of many enzymes as well as blood cells.

Fish require seven main minerals (calcium, phosphorus, potassium, sodium, chlorine, magnesium, and sulfur) and 15 other trace minerals. These can be supplied by diet but can also be directly absorbed from the water through their skin and gills. Supplementing of vitamins and minerals can be done according to the requirements of each species. The production of feed requires a fine balance of all of the nutrient components mentioned above (protein, lipids, carbohydrates, vitamins, minerals, and total energy). An unbalanced feed will cause reduced growth, nutritional disorders, illness and, eventually, higher production costs.

Fishmeal is regarded as the best protein source for aquatic animals because of its very high protein content and it has balanced EAAs. However, it is an increasingly expensive ingredient, with concerns regarding sustainability. Moreover, fishmeal is not

TABLE 12. **Mineral requirements of tilapia.**

MINERAL	SPECIES	REQUIREMENTS (MG/KG)
Calcium	O. aureus	7000
Phosphorus	O. niloticus	< 9000
	O. niloticus	4600
	O. niloticus x O. aureus	7000
	O. aureus	5000
Potassium	O. niloticus x O. aureus	2100-3300
Magnesium	O. aureus	500
	O. niloticus	600-800
Zinc	O. niloticus	30
	O. niloticus	79
	O. aureus	20
Manganese	O. nilolicus	12
Iron	O. niloticus	60
Chromium	O. niloticus x O. aureus	139.6

always available. Proteins of plant origin can adequately replace fishmeal; however, they should undergo physical (de-hulling, grinding) and thermal processes to improve their digestibility. Plant ingredients are, in fact, high in anti-nutritional factors that interfere with the digestion and the assimilation of nutrients by the animals, which eventually results in poor fish growth and performance.

The size of the pellets should be about 20–30 percent of the fish's mouth in order to facilitate ingestion and avoid any loss. If the pellets are too small, fish exert more energy to consume them; if too large, the fish will be unable to eat. A recommended pellet size for fish below 50 g is 2 mm, while 4 mm is ideal for pre-adults of more than 50 g.

The use of any raw ingredient of animal origin (fish offal, blood meal, insects, etc.) should be preventively heat treated to prevent any microbial contamination of the aquaculture system.

Fiber – Tilapia (Do not exceed 5%)

Fiber is usually considered indigestible, as tilapia do not possess the required enzymes for fiber digestion. For this reason, and to attain maximum growth, crude fiber levels in tilapia diets should not exceed 5%.

TABLE 13. **Vitamin requirements and deficiency symptoms of tilapia.**

VITAMIN	FISH SPECIES	REQUIREMENT (MG/KG)	DEFICIENCY SYMPTOMS
Niacin	O. n x O. a	26 121	Haemorrhage. deformed snout Gill and skin oedema. fin and mouth lesions
Biotin	O. n x O. a	0.06	
Choline	O. n x O. a	1000	
Pantolhenic Acid	O. aureus	10	Low growth, haemormage. anaemia, sluggishness, hilt! mor1ality, hyperplasia of epithelial cells of gill lamellae
Thiamin	O. m x O. n	2.5	I ow growth and low food efficiency, low hacmatocrit
Riboflavin	O. aureus	6	Anorexia, low growth, hilt! rnort.ality, fin erosion, loss of body cdour, cataract, dwarfism
Aiboflavin	O. m x O. n	5	
Pyridoxine	O. n x O. a	1.7-9.5 15-16.5	Low growth, high mortality, abnormal neurological signs, anorexia, convulsions, caudal fin erosion, mouth lesion
VitaminC	O. aureus O. n x O. a O. nil0ticus O. n x O. a O. n x O. a O. n x O. a O. spilurus	50 79 420 41-48 37-42 63.4 11- 200	
D	O. n x O. a	374.S IU	
E	O. aureus	10 25	Low growth and FCR, skin haemorrhage, muscle dystrophy, impaired erythropoiesis, oeroid in liver and spleen, abnoonal skin coloration
E	O. nil0ticus	50-100 500	
E	O. n x O. a	42-44 60-66	
A	O. nil0ticus	5000 IU	Low growth, resdess, abnormal movement, blindness, exophthalmia, haemorrhage, pot-belly syndrome, reduced mucus secretion, high mortality
Inositol	O. n x O. a	400	

Homemade Fish Feed for Omnivorous and Herbivorous Fish

Two simple recipes for a balanced fish feed containing 30 percent of crude protein (CP) are provided in tables 14 and 15. The lists of the ingredients for each diet are expressed in weight (kilograms), enough to make 10 kg of feed. The first formula is made with proteins of vegetable origin, mainly soybean meal (see table 14). The second formula is mainly made with fishmeal (see table 15).

Step-by-Step Preparation of Homemade Fish Feed

1. Gather the utensils noted in table 16.
2. Gather the ingredients shown in tables 14 and 15. Purchase previously dried and defatted soybean meal, corn meal, and wheat flour. If these meals are unavailable, obtain whole soybeans, corn kernels, and wheat berries. These would need to be dried, de-hulled, and ground. Whole soybeans need to be toasted at 240°F (120 °C) for 1–2 minutes.
3. Weigh each ingredient following the quantities shown in the recipes above.
4. Add the dry ingredients (flours and meals) and mix thoroughly for 5–10 minutes until the mix becomes homogeneous.
5. Add the vitamin and mineral premix to the dry ingredients and mix thoroughly for another 5 minutes. Make sure that the vitamins and minerals are evenly distributed throughout the whole mixture.
6. Add the soybean oil and continue to mix for 3–5 minutes.
7. Add water to the mixture to obtain a soft, but not sticky, dough.
8. Steam-cook the dough to cause gelatinization.
9. Remove the dough, divide into manageable pieces, and pass them through the meat mincer/pasta maker to obtain spaghetti-like strips. The mincer disc should be chosen according to the desired pellet size.
10. Dry the dough by spreading the strips out on aluminum trays. If available, dry the feed strips in an electric oven at a temperature of 140–185 °F 60–85 °C for 10–30 minutes to gelatinize starch. Check the strips regularly to avoid any burn.
11. Crumble the dry strips. Break or cut the feed on the tray with the fingers into smaller pieces. Try to make the pel- lets the same size. Avoid excessive pellet manipulation to prevent crumbling. Pellets can be sieved and separated in batches of homogeneous size with proper mesh sizes.
12. Store the feed.

Storing Homemade Fish Feed

Once prepared, the best way to store fish feed is to put pellets into an airtight container soon after being dried and broken apart. Containers must be kept in a cool, dry, dark and ventilated place, away from pests. Keeping pellets at low levels of moisture (less than 10 percent) prevents them becoming moldy and developing toxic mycotoxins. Depending on the temperature, the pellets can be stored for as long as two months.

Another way to keep pellets for long periods is to close them in a plastic container and store them in the refrigerator. Feed kept in this way can last for more than one year.

Feed must be used on a "first in, first out" basis. Avoid using any feed showing signs of decay or mold, as this could be fatal for fish.

TABLE 14. **Recipe for 10 kg of fish feed using vegetable-based protein, including proximate analysis**

FEED INGREDIENTS	WEIGHT (kg)	% OF TOTAL FEED	PROXIMATE ANALYSIS	%
Corn meal	1.0	10	Dry matter	91.2
Wheat flour	1.0	10	Crude protein	30.0
Soybean meal	6.7	67.2	Crude fat	14.2
Soybean oil	0.2	2	Crude fiber	4.8
Wheat bran	0.7	7.8	Ash	4.6
Vitamin and mineral premix	0.3	3	Nitrogen-free extract (NFE)	28.3
Total amount	10.0	100	-	-

TABLE 15. **Recipe for 10 kg of fish feed using animal-based protein, including proximate analysis**

FEED INGREDIENTS	WEIGHT (kg)	% OF TOTAL FEED	PROXIMATE ANALYSIS	%
Corn meal	1.0	10	Dry matter	90.9
Wheat flour	4.0	40	Crude protein	30.0
Soybean meal	1.5	15	Crude fat	10.5
Soybean oil	0.2	2	Crude fiber	2.1
Fishmeal	3.0	30	Ash	8.3
Vitamin and mineral premix	0.3	3	Nitrogen-free extract (NFE)	34.5
Total amount	10.0	100	-	-

TABLE 16. **Utensils**

COMPONENT	QUANTITY	SPECIFICATION
Weighing scale	1	Capacity 2–6 lbs. (1–3 kg), Divisions in ounces or grams
Grinder	1	Electric coffee-type grinder
Metal sieve	1	5–10 U.S. Sieve Size, 0.2–0.4 cm (2–4 mm) mesh
Mixing bucket	1	Capacity 3-gallon (10 liters)
Plastic bowl	1	Capacity 2-quart (2 liters)
Meat mincer / pasta maker	1	Manual or electric
Mixing spoon	1	Large size
Aluminum baking tray	10	Large baking tray

Feeding Summary

- A good rule of thumb is to feed your fish as much as they will eat in 5 to 10 minutes, 1 — 3 times per day. An adult fish will eat approximately 1 percent of its bodyweight per day. Fish fry (babies) will eat as much as 7 percent. Be sure that your fish are being fed enough. However, be cognizant of the fact that over feeding fish will negatively affect water quality, is wasteful, and is an unnecessary increase in cost.
- If your fish are not eating as they should, it is a good indication that they are stressed or unhealthy. Some factors that may result in fish not eating as they should:
    ◊ Living in conditions that are outside of their optimal temperature range.
    ◊ Water quality issues: Improper pH range, too much ammonia in the system, inadequate dissolved oxygen.
    ◊ Loud or irritating noises and vibrations.
    ◊ Direct lighting upon the fish tank.

PART III : FISH

PART III

Components Used in Aquaculture

PART III : COMPONENTS USED IN AQUACULTURE

CHAPTER 6

Equipment & Component Overview

Overview of Aquaculture Equipment and Components

Following is a list of equipment that is available for a Tilapia farming system. Some components are necessary, whereas other items are optional, or may not even be applicable to your particular set-up. The list is provided for consideration. All of these items are available via the internet, and many can be obtained through your local home improvement store. Prices vary greatly, depending on the quality of parts and materials used and how sophisticated you want to go, i.e., automation vs. manual.

Only use food-grade plastics and materials typically used for potable water such as PVC. Some plastics will degrade from weather and break down or leak chemicals into your system. Avoid using copper or other metals in any part of the system exposed to water, as they can negatively affect the fish.

Some aquaculture operators use recycled items for system parts. This is a great way to lower costs; however, it is important to refrain from using materials previously used with chemicals or when their prior use is unknown. Even used parts with marine grade paint or galvanized coatings can leak chemicals and oxidized metal into your system. Inspect and thoroughly rinse all parts thoroughly before integrating them into your system. For additional protection certain parts can be encased with a safe marine grade epoxy coating.

Below is a list of items used in a Tilapia farm operation. These items will be addressed in much greater detail in the following chapters, with the most prominent components (pumps, grow media, plumbing, fish tanks, liners, etc.) each having an entire chapter dedicated to providing more comprehensive information.

Plumbing

Plumbing supplies which may include piping, elbows, bends, sleeves, pipe joint compound, gaskets, flares, reducers, tees, Y-joint, etc. will be needed. In 'Chapter 8: Plumbing', all aspects of plumbing will be addressed in detail.

Float Switches

Float switches are inexpensive devices used to control the pump depending on the water level (see figure 25). If the water level in the fish or sump tank falls below a certain height, the switch will turn off the pump. This prevents the pump from pumping all of the water out of the tank. Similarly, float switches can be used to fill the aquaponic system with water

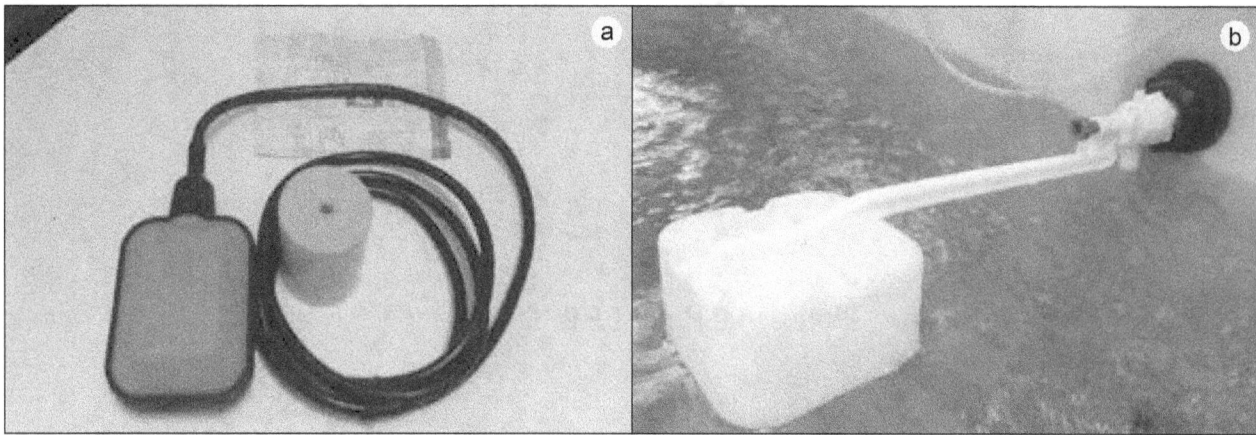

FIGURE 25. Float switch controlling a water pump (a) and a ballcock and float valve controlling the water main (b)

from a hose or water main. A float switch similar to a toilet ballcock and valve can ensure the water level never falls below a certain point. It is particularly important to know that in certain types of loss-of-water events, such as a bro- ken pipe, a float switch can actually make the flooding much worse, and this needs to be carefully considered in indoor applications and other situations where flooding could cause significant property damage or electrical shock dangers.

Hi-Low Water Level Sump Pump Controller (Dual Float Switch)

A Hi-Lo Pump Controller is a dual float device with a universal switch that works with all types of sump pumps and utility pumps. Its two sensors give you complete control of where your pump turns on and off. Using this type of controller with the right utility pump allows you to set the turn on level as low as 1/2" of water and turn off as low as 1/8" of water. When used with a sump pump the controller enables you to adjust the turn on and turn off levels to get the longest run time for the pump, which saves energy and lengthens the service life of the pump.

One example of a Hi-Lo pump controller switch is the HC6000 by HydroCheck. It is currently available on Amazon. com for $83 (USD, year 2021), and has excellent reviews. This particular make/model Hi-Lo pump controller switch is presented as an example and hopefully a helpful starting point (resource) for the reader (see figure 26). There are also other Hi-Lo switches on the market with various features and degrees of quality, and cost.

Water Detector and Hi-Low Water Alarms (Optional)

There is a wide variety of water alarms available with a range of features, quality, and cost. They are available with battery, solar, and/or AC/DC power supply. Most units under $100 have sound alarms, but some also have a light 'on' alarm. More expensive units can be obtained that make a phone call, send an email, and/or a text message alert.

Water detection and hi-low water alarm devices are not mandatory in aquaponics but can provide the operator with a higher level of protection and peace of mind. They can also help prevent a disaster, saving the aquaponics operator from potential system-related loss (fish, pump burn out, plant) and/or property damage (flooding).

For instance, a water detector sensor alarm could be used where water could damage the surrounding environment (i.e., indoor applications). A hi-low water alarm can alert the operator 'immediately' of any water balance problems. Discovering the problem later during a routine operational check may be too late.

Again, these devices are entirely optional. They range in cost from under $10 to over a $1,000 (USD, year 2021). This expense needs to be weighed against risk factors, budget, and peace of mind.

- Works with all types of sump pumps and utility pumps
- Precise control of turn on and turn off levels
- Wide control range from 1/2" to 20' (Not timed based like other dual-float switches)
- Small profile fits in small or crowded spaces
- Sensors not affected by minerals and debris in water — Never has to be cleaned
- Part Number HC6000
- Product Dimensions 2.8 x 3.8 x 2.5 inches
- Voltage 120 volts; Wattage 2.4 watts; Amperage Capacity 14 A
- Cord Length 12 Feet
- UL Certified
- HC6000 is for indoor applications.
NOTE: For outdoor applications refer to HC6100 by HydroCheck.

FIGURE 26.

Other Monitoring Alarms

Automatic monitoring devices for water temperature, water level, pumps, pH, blower, lights, air temperature, dissolved oxygen, or the entire system and system environment are available. Alarms can even call, text message or email you the status of conditions and alert you if there is a problem.

Water Quality Testing

Monitoring water quality in aquaponics is an important procedure. There are diverse water quality test kits and meters available, manual and automatic, in a wide price range. Water quality testing devices can be acquired which will monitor just one, several or all of the following components. Although all of these water parameters should be tested, budgetary constraints may determine how many tools are used to gather this data.
- pH
- Ammonia
- Water Temperature
- Dissolved Oxygen Inline Water Heater
- Allows the operator to establish the desired water temperature. Most devices are fully equipped with a digital control mechanism that will provide precise all-season temperature regulation. They also come with a flow switch and safety thermostat and are easy to install with the option of water flow from either direction.

Ammonia Nitrogen Test Kit

Used for determining ammonia and nitrogen concentrations in water.

Water Hardness Test Kit

A portable testing kit to measure total hardness, calcium, and magnesium hardness in water.

Fish Tank Shape

Avoid tanks that have hollow pockets as they will collect fish waste that will go anaerobic and cause water quality problems. Also ensure that there is adequate circulation delivering dissolved oxygen throughout the tank.

Liner

A liner is an impermeable geotextile used for water retention commonly referred to as pond liner, water garden liner, greenhouse cover material, hydroponics pond liner, aquaponics bed liner, and polyethylene tarp material.

Materials include LDPE (Low Density Polyethylene), PVC (Poly Vinyl Chloride), PVC with internal reinforcement, and HDPE (High Density Polyethylene). HDPE is frequently noted as the best, but it is the most expensive and hardest to work with because of its rigid nature. LDPE is most preferred by small to mid-size operators. Avoid EPDM as it emits toxins harmful to the beneficial bacteria in your system, and the toxins will eventually make their way into your food supply. Also avoid vinyl liners (they are too stretchy) and any product that does not specifically identify what it is made of.

The liner thickness should range anywhere from 20 to 40 mils. Obviously the thicker the better in terms of protection against tears. However, the thicker the liner the more difficult it is to handle and work with during installation; thicker will also be more expensive.

The ideal liner will be UV resistant, fiber reinforced, have one side white for easier installation, food grade quality, and thick enough to not tear easily during installation or regular use. See 'Chapter 9: Liner Material' for a comprehensive examination of liners.

Pumps

The correct pump for your systems is important. Some ill-informed people simply purchase a pump at the local home improvement store or order one online before gaining an understanding of their system's parameters. This is a big mistake. Please do not buy your pump this way. It is rare to find a sales representative or a retailer that knows aquaculture (growing fish). A pump needs to be selected based upon compatibility of pump specifications to the system parameters, such as how much water is in your system, how high the pump needs to raise the water (head), desired flow rate, and whether or not a siphon or timer system is being used. If you desire to have your system off the grid, then efficiency becomes a very critical factor as well. Even if you have your system on the grid, efficiency should be considered to minimize operating costs. Refer to 'Chapter 11: Pumps & Choosing the Right Pump' for more specific and detailed helpful information on pumps.

Aeration

Aeration devices are sometime needed, or just included for additional benefit, in order to raise dissolved oxygen levels. Oxygen is introduced into the aquaponics system naturally through the plant life, waves, cascades, and other water turbulence methods, such as fountains and waterfalls. If water turbulence and natural aeration methods, such as with certain types of plants, are insufficient then an aeration device is necessary. Aeration devices are also used if dead spaces in a tank are a problem such as bottom corners of square tanks, end of a long run, etc. Mechanical devices include air pumps, tubing, oxygen injectors and oxygen diffusers.

Timing Controllers

Timing Controllers enable you to run your system according to your preferences and needs. They can be set up to work a pump, multiple pumps, water heaters, lighting, fish feeding, humidity, and anything else you desire to automate. Having components on an automatic timer controller is a nice convenience. It allows the operator to maintain the system on a consistent schedule and better enables one to make clearly defined adjustments to various system elements (flow rate, water temperature, feeding schedule, lighting preferences, etc.). Timing controllers come with a wide range of features and in many price ranges.

Ambient Lighting

Work lighting is also an important consideration. Just remember not to have any direct lighting impacting the fish tank.

Backup Energy

In case of an extended power outage, it is essential to have a back-up energy source. Keep in mind that if your power is out for several days, others in your community will also most likely be without power, resulting in a run on generators from area rental companies and home improvement stores. Fish will perish quickly as oxygen becomes insufficient and waste begins to accumulate.

An electric generator is a device that converts mechanical energy to electrical energy, typically via burning some type of fossil fuel (gas, diesel, propane, etc.). Typically, the only items that must be provided power through an emergency event are pumps, aerators, and heaters. The remaining aquaculture components being without power through an extended outage will typically not result in catastrophic problems, so long as the operator manually manages those systems and parameters properly during the outage.

For small to mid-size systems, a pump can simply be plugged in to a power inventor box, which is in turn connected to any DC power source (typically just a 12-volt automobile battery). The AC output voltage of a power inverter device is often the same as the standard power line voltage, such as household 120AC. This allows the inverter to power numerous types of equipment designed to operate off the standard line power. Most often, if it is just a matter of a few hours, survival can be maintained by just keeping the pump in operation.

Alternative Energy

Alternative energy for aquaculture is referred to as any energy source that provides off-the-grid power supply. This back-up power can be used in case there is a disruption of service of the main public utility supply, or as a primary means in which to power the electrical components associated with an aquaculture system (i.g., pumps, lighting, heat, fans, etc.). Following is an example of one back-up alternative power supply which is relatively affordable.

Alternative Energy Supply Source (one example)
Sunforce 50048 60W Solar Charging Kit
Available on online for approximately $329 (USA, year 2020)

Features:
- Amorphous solar charging kit provides up to 60 watts of clean, free, renewable power
- Designed for back-up and remote power use
- Weatherproof, durable solar panels.
- Built-in blocking diode helps protect against battery discharge at night
- Complete kit includes four 15W amorphous solar panels, a PVC mounting frame, a 7-amp charge controller, 200-watt inverter, and wiring/connection cables

FIGURE 27.

PART III : COMPONENTS USED IN AQUACULTURE

CHAPTER 7

Fish Tanks and Ponds

Fish Tank Shape

Although any shape of fish tank will work, round tanks with flat bottoms are recommended. Square tanks with flat bottoms are perfectly acceptable but require more active solid-waste removal. Tank shape greatly affects water circulation. There is a risk to having tank with poor circulation. Artistically shaped tanks with non-geometric shapes (featuring many curves and bends, such as a Jacuzzi) can create dead spots in the water with no circulation. These areas can gather wastes and create anoxic, dangerous conditions for the fish. If an odd-shaped tank is to be used, it may be necessary to add water pumps or air pumps to ensure proper circulation and remove the solids. It is also important to choose a tank to fit the characteristics of the aquatic species reared because many species of bottom dwelling fish show better growth and less stress with adequate horizontal space.

As mentioned above, the fish tank can be of almost any shape, but the most commonly used are circular. The round shape allows water to circulate uniformly and transports solid wastes towards the center of the tank by centripetal force. A round tank is structurally stronger than any other shape, so given that the tank may be holding 250 plus gallons of water, the round shape allows for better structural integrity and less reinforcement. Therefore, round tanks don't have to be as heavy and bulky as square or rectangular

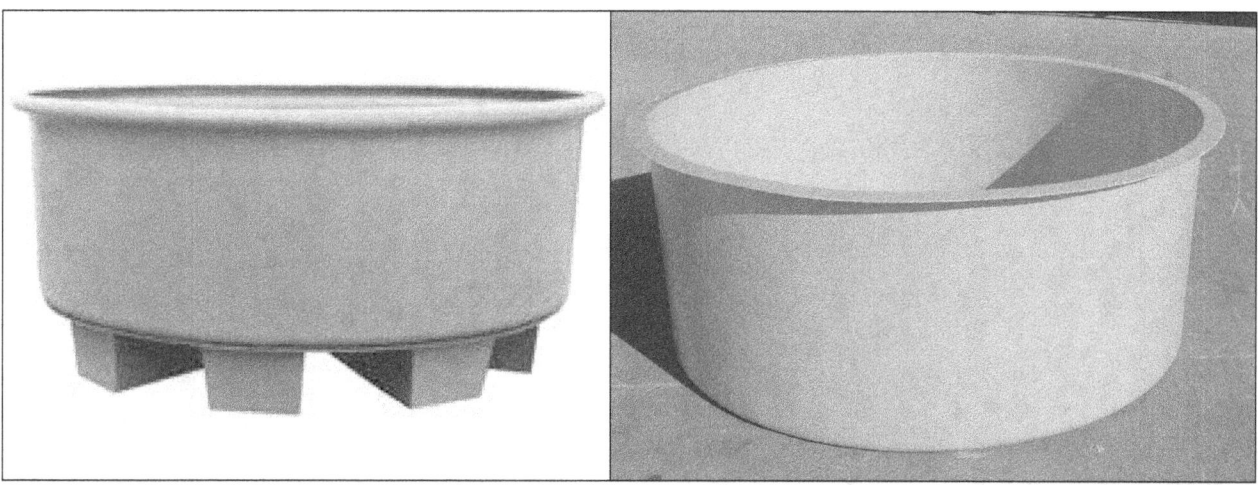

FIGURE 28. Round fiberglass tanks

FIGURE 29.

shaped tanks. As a result, round tanks, such as the fiberglass tanks shown in the two images (figure 27), are usually the least expensive tanks.

Round tanks have several distinct benefits for water quality and fish health, such as:

- Allows for better water circulation.
- Water circulation prevents thermal layering and thus helps to improve water quality.
- A good flow of water provides the fish with a current to swim against, which promotes health and muscle growth. This in turns makes raising fish for food better for consumption.
- Fish appear to naturally enjoy some moving water.
- Water movement aids in the exchange of gases between the air and the water (CO_2 exiting the fish tank, and oxygen entering the fish tank). Surface water constantly changing dramatically increases the rate of oxygen exchange.
- Increased oxygen through water movement benefits not only the fish, but also the de-nitrifying bacteria in the system, as well as inhibiting harmful bacteria that thrive in an anaerobic environment.
- In a rounded tank, solid waste has a tendency to gravitate toward the bottom-center of the tank. Strategic placement of the pump at this point will enable the collection and transport of this solid waste to the grow bed.

NOTE: I highly recommend that you avoid bolt-up fiberglass tanks, similar to the ones in figure 28. I have had lots of experience with these types of tanks, as well as many other types of fish tanks. Bolt-up fiberglass tanks are highly problematic. There is a strong probability for leakage with bolt up tanks (eventually, if not initially). I have seen bolt up tanks leak way too many times in my career. I will never use or recommend a bolt up tank. Even if obtained for free, they will cost you time, and most likely product, when you have to eventually address a leak. Lastly, I believe bolt up tanks can be a safety hazard as well, should leaked water make its way to an electrical source (e.g., extension cord on the ground, etc.).

Square or rectangular fish tanks are also common and are especially useful if there is limited space. Care should be taken to ensure there is adequate water circulation so dead air pockets (low dissolved oxygen levels) do not form in the corners. Square tanks with rounded corners will assist in the water movement.

FIGURE 30. Rectangular and square tanks.

IBC and Macrobin Tanks

IBC (Intermediate Bulk Carrier) totes (or high quality, food grade bulk shipping containers such as Macrobins) are commonly used for fish tanks. A used or reconditioned 275 or 330-gallon food grade IBC tote works great and can be acquired at a relatively low cost. Most standard size micro-bins can hold about 400 gallons of water.

Liner Tanks

Another practical and relatively inexpensive type of tank that works well is a DIY frame (wood, brick, concrete block, earthen, etc.) lined with a LDPE (Low Density Polyethylene) or HDPE (High Density Polyethylene) liner. The liner should range anywhere from 20 to 40 mils in thickness. Obviously, the thicker the liner, the better in terms of protection against tears. However, the thicker the liner the more difficult it is to handle and work with during installation, and the more expensive it will be.

The ideal liner would be UV resistant, fiber reinforced, have one side white for easier installation, be of food-grade quality, and be thick enough so as to not tear easily during installation or regular use. However, a liner that is made of LDPE or HDPE material, is 20 to 40 mils in thickness, and UV resistant is certainly satisfactory.

FIGURE 31. IBC (Intermediate Bulk Carrier) Tote

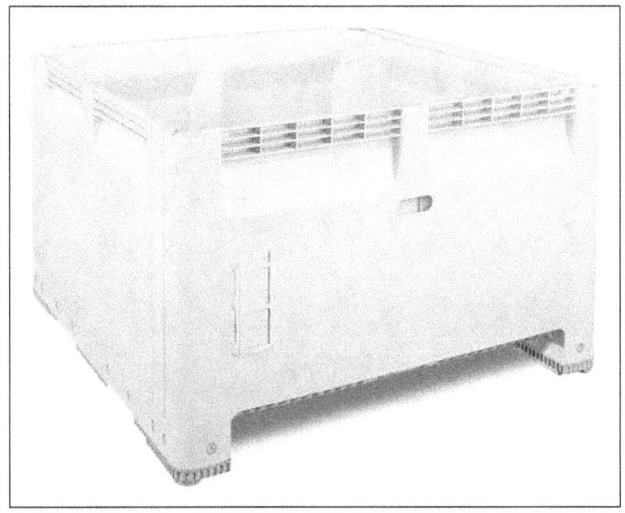

FIGURE 32. A Macrobin

FIGURE 33. Liner tanks

Fish In and Predators Out

Regardless of what type of tank is used, it is important to ensure that small children and predators cannot enter the tank, and that fish cannot exit the tank. To keep fish in the tank, the water level can be lowered to approximately eight inches below the rim (this works for most species). However, doing so greatly decreases valuable volume capacity. A better method is to install netting or a fence around or over the fish tank. Keep in mind that it is also important to have quick access to the fish tank to check on the water quality, and for maintenance purposes.

I worked with a large aquaculture farm in California that raised several different types of fish in both ponds and 50-foot diameter tanks. One species raised in the tanks was largemouth bass. Although the surface water level was about a foot below the rim of the tanks, they had to keep a net stretched tightly over the bass tanks, because during feeding (when they dispersed feed on the surface of the water) the fish would get so excited--the water became extremely turbulent--the fish would literally jump out of the tank if it were not for the net.

Multipurpose Fish Tank (Pond)

Planned correctly, a fish tank can serve several valuable purposes. The benefits of each should not be overlooked or taken lightly. Using a fish tank for more than just containing fish can significantly increase production, profits, and lower cost. And although I am using the 'tank' term, the same principles apply to a pond or any vessel used to raise fish. Furthermore, a fish tank's value increases when each of the following items are implemented.

- **Fish tank used for radiating heat and/or the cooling of the greenhouse.** Radiation is the process of heat transfer resulting from the temperature difference between elements. Standing next to a hot fire (or standing outside on a hot sunny day), you can feel radiation heating the

FIGURE 34. Net over fish tank

FIGURE 35. fence around fish tank

surface of your skin. Similarly, in a greenhouse, a fish tank radiates energy to the surrounding environment. In the summer, the tank helps cool the greenhouse, and in the winter the tank helps generate heat. This process helps to lower heating and cooling utility costs and helps maintain a more consistent temperature balance within the greenhouse.

- **Fish tank used with an aquaponic DWC raft.** A D.W.C. plant raft on the water surface of the fish tank provides shade and security for the fish, while at the same time allows for the growing of certain beneficial plants. Besides providing you plant product without sacrificing space, the raft helps the fish feel more secure, and as a result, lowers fish stress thereby improving their health. Just be sure to allow some spacing between the edge of the plant raft and the side of the tank for fish care and tank maintenance.

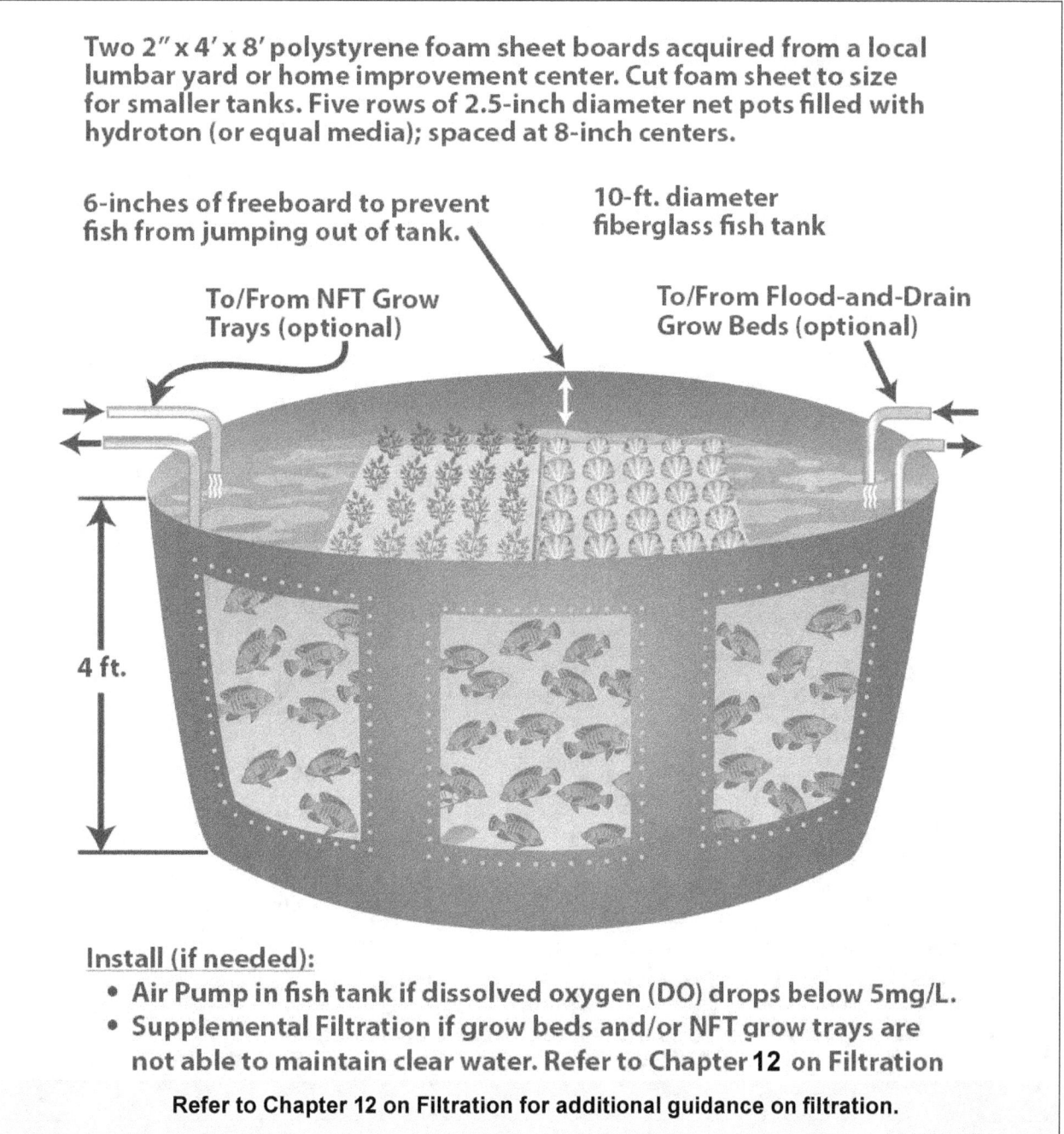

FIGURE 36. Fish tank integrated into an aquaponic system, providing multiple uses. The plant raft serves as a Deep Water Culture (DWC) for plants and fish shading. Grow beds elsewhere can be utilized for plants and help filter the water.

- **Fish tank used for multiple types of aquaponic systems.** Most aquaponic resources in the marketplace refer to the three different types of systems (D.W.C., N.F.T., and Flood-and-Drain). However, if planned properly, a system can be set up to be more than just one system, thus providing much greater flexibility and product yield.

Figures 35 and 36 show how a fish tank/pond can be utilized to provide "multiple-use" options, and thereby maximizing production without sacrificing space.

NOTE: Introducing increased oxygen into the water, via aeration tubes, not only helps keep the plants healthy, but it also accelerates plant growth.

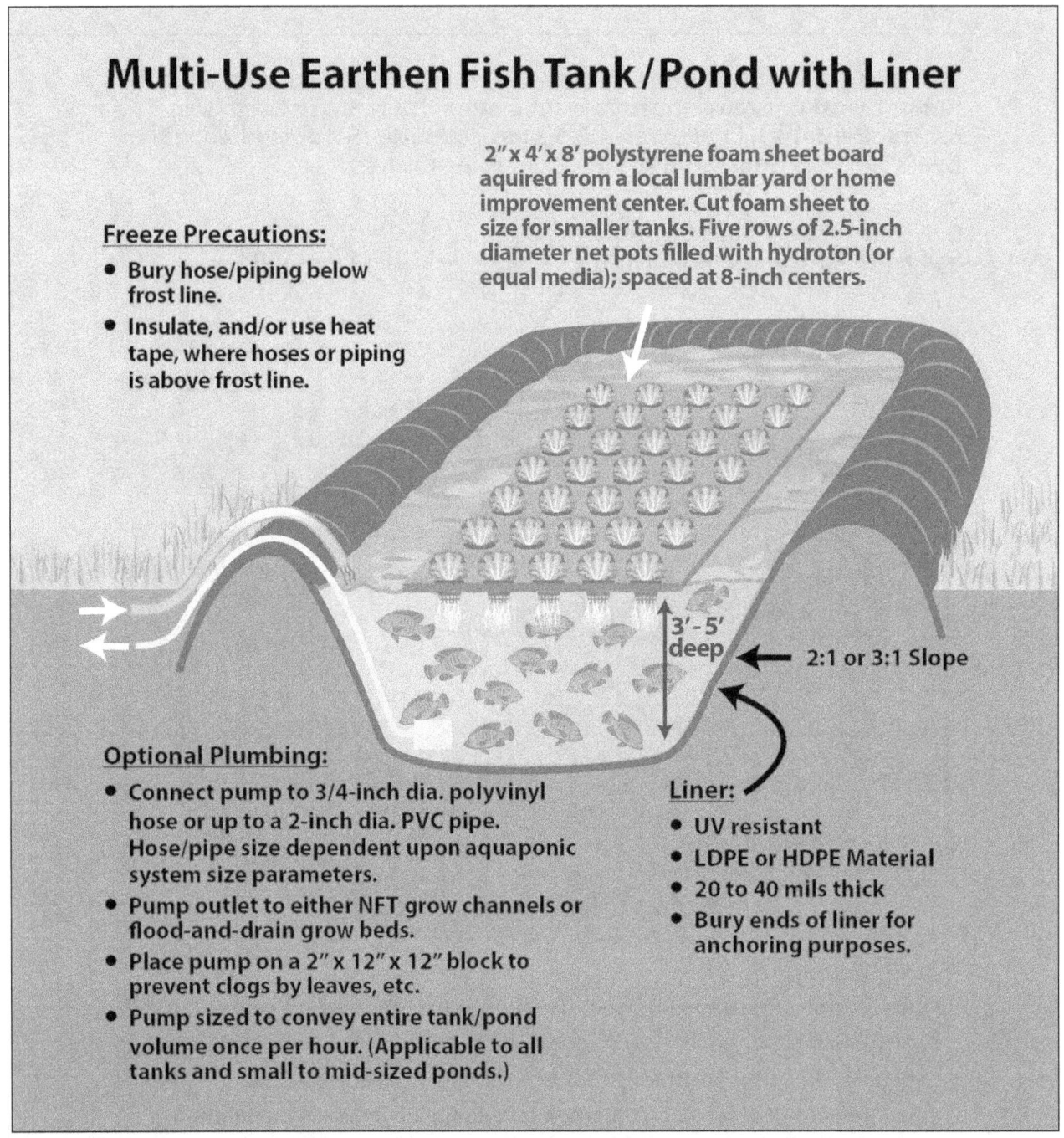

FIGURE 37. *Fish pond serving multiple aquaponic uses (DWC, plus optional plumbing for NFT grow channels and/or Flood-and-Drain media-beds).* **NOTE:** *In some circumstances an air pump may be necessary in order to deliver additional oxygen to the DWC plant roots.*

Birds of Prey

As mentioned in the introduction of this book, I served as the Engineering Manager over four large aquaculture farms, which produced 80 percent of the Nation's caviar and seven-plus tons of meat annually, which we also shipped meat and caviar all over the world. True caviar is harvested from sturgeon when they are ten-plus years old. In my case, we raised white sturgeon fish from egg to harvest. By the time they were old enough to produce caviar, they typically weighed anywhere from 125 to 300 lbs. (four to seven-plus feet in length). One of our farms was inside a steel building, and the other three farms were outdoors. All four farms had a water recirculation system, and the fish were raised in round tanks; most tanks ranged in size from 20 to 50 feet in diameter, with a depth of four to six feet. However, the larger tanks were ten feet deep. These tanks held a lot of fish. We had to keep all outdoor tanks with fish under about two years of age (less than 18-inches in length) covered with a net or shade cloth to prevent hawks and eagles from stealing them. The tanks were checked a least three times a day (feedings, maintenance, etc.), so employees had to regularly partially uncover the tanks to gain access. There were hundreds of tanks.

On rare occasions, an employee would fail to tightly seal the tank cover, after which we would sometimes find the remains of a partially eaten fish where a bird of prey took advantage of the opportunity for an easy meal. If you live in a location where hawks and eagles are common, you may need to install a cover over your tank, if you find these awesome birds of prey are stealing your fish. It is doubtful that a scarecrow would work, as these birds will try to steal a fish off of a fishing line should they encounter a fisherman reeling in his catch.

FIGURE 38. *A hawk with his catch.*

FIGURE 39. *A shade covers over a fish tank helps regulate water temperature, shields fish from direct sunlight, and provides protection against birds of prey taking fish.*

FIGURE 40. *An eagle stealing a fish off of a fishing line.*

Fish Tank Materials

Fish tanks are made via many different methods, and with a variety of materials. Some of the most common are food-grade plastic, fiberglass, and HDPE liner. However, concrete tanks and earthen structures are also used. All have their advantages and disadvantages. For practicality, in consideration of cost and ease of installation, plastic, fiberglass and HDPE liner materials make the most sense if the tank is going to be located on grade. The materials for these types of tanks are light weight, easy to handle, and readily available throughout most of the world. Concrete and earthen tanks are most practical if the tank is below grade. Some creative aquaculture enthusiasts have even successfully converted swimming pools into viable aquaculture systems.

Wooden frames (or any other type of material for that matter) can also be built and turned waterproof via a HDPE liner, fiberglass, or an epoxy-based substance. These methods are certainly acceptable, but care needs to be taken to ensure that the frame has the structural integrity to withstand the lateral pressures, downward force, and stresses of the water weight against the structure; and the potential of several adults leaning against the tank. Water alone is extremely heavy by volume. Add to that a grown person leaning against the tank, and it is easy to understand that the tank must be fabricated using strong materials and heavy-duty construction methods.

Ponds are also used in farming Tilapia. With ponds extra should be taken as an outside pond may contain and spread undesirable bacteria and/or sediment throughout the aquaculture system via the recirculating pumped water. This would compromise the water quality of the system and, at the least, require more frequent cleanings.

Concrete needs to be carefully considered and managed, as it can negatively impact water composition by causing imbalances in the pH. As water interacts with concrete, the water can dissolve various minerals present in the hardened cement paste or within the concrete aggregates. If the solution is unsaturated, dissolved ions such as calcium (Ca^{2+}) are leached out and can be detrimental to the aquaculture system as a whole. If concrete is used, it should be sealed with a food-safe, commercially available sealant

If constructing an aquaculture system with a previously used tank or materials, care needs to be taken to ensure that the tank is food grade, materials don't have any toxic residues, and there are no elements associated with what is being used that may be harmful to fish, plants, and human consumption. If the tank will be outdoors, it should also be of material that will not breakdown easily via UV light (UV resistant). If you are unsure about the material or tank, then it would be prudent to pass on it as potential problems are not worth the risk, could end up being a costly mistake, would be a major drain on your valuable time, and negatively impact on your peace of mind.

Ideally, the fish tank will be of an opaque material (not transparent), as direct sunlight will encourage algae growth. Algae is detrimental to an aquaculture system (primarily because algae are a prolific grower that also needs oxygen to multiply, thus depleting your system of the available oxygen needed by your fish, bacteria, and plants).

A window on the side of a tank can bring much pleasure, and also serves to make monitoring water quality and fish health much easier. It also makes aquacultures that much more fun for both adults and children. However, it is best if such a window is not located on a side of the tank that is exposed to direct sunlight, but if so, install a curtain so the sunlight can be blocked when the window is not in use.

Small aquariums that one finds in homes or businesses are able to use standard glass because they are micromanaged and have specific aquarium filters which address algae. Even so, a small aquarium can develop algae if in direct sunlight or if the filter is inadequate.

Fish Tank Size

There are several different issues that must be considered pertaining to fish tank size selection. Below is a summarized list:

Objectives, goals, and long-term plan.

What is your purpose for getting into aquaculture? What are your immediate and long-term production goals (e.g., amount of fish you want to harvest each year)? Do you want to install your system and be done with it, or do you want to have the ability to enlarge it over time? Is your objective to just partially supplement your grocery store food with healthy, home-grown food, grow enough for yourself and/or others in your circle, or also grow enough to barter and/or sell for a profit?

Location, facilities, and available space.

Will your tank be indoors or outdoors? A fish tank in a greenhouse works great because it helps maintain the temperature in the greenhouse. Will you need a greenhouse or green room to grow vegetables in the winter? How much space do you have available for both the fish tank and plants? Will your fish tank require temperature control during all or part of the year?

Start-up and operational budget (feed cost, utilities, etc.)

The larger the system, the greater the start-up and operational cost. How much are you willing to invest in creating your system? What is your budget for feeding the fish, and potentially pay for increased utility cost (for temperature control and lighting)?

Maintenance and operational resources (e.g., your time, available assistance, etc.)

The larger the system, the more time and attention will be required. It is great to have big ambitions, but if you don't have the time or resources to keep up with it, success will be hindered. Sometimes it is better to start off small and then grow in phases.

Regardless of the type of system or size of the fish tank, a certain level of water quality must exist. Adequate filtration must occur, whether it is through aquaponic grow beds, a bio-filter, mechanical filtration, or a combination of these filtration methods. If growing fish in a pond, will the fish be negatively impacted by runoff surface water contaminants entering the pond, extreme fluctuations in water levels, or predators (including poachers)? Water quality and specific details of system design are addressed later in this book

Redundancy

The optimal planned aquaculture system will have redundancy integrated within, where possible. Using a minimum of two fish tanks, two pumps, and two filtration systems. Redundancy offers many advantages when it comes to addressing emergencies, harvesting operations, and maintenance issues.

Fish Tank Placement

A common perspective of aquaculture is that of having the fish tank downstream of the filtration system. Although there is nothing wrong with this layout, it is by no means the only option. As a matter of fact, sometimes it is a big advantage to have the fish tank outdoors.

The fish tank is the heaviest component in the aquaculture system. Each gallon of water weighs 8.34 lbs., which means a 300-gallon fish tank would weigh about 2,502 pounds (slightly less because of the fish). Since the fish tank will be too difficult to be move once it is filled, it is important to carefully consider its location. Furthermore, the fish tank should be located within or on the ground, or on solid flooring, since a structure to support all the weight would have to be enormous and very costly. In short, due diligence in determining fish tank location is paramount.

Fish Tank Safety Considerations

Regardless of the tank system being implemented or its size, remember that safety is a priority. If you have small children on your property then fencing, nets, and/or locked doors should be intact to prevent children from potentially drowning in the fish tank.

Water in contact with electrical components, electrical outlets, breaker boxes, and extension cords combine for an extremely dangerous situation. All electrical items should be securely protected from water and potential fish tank leaks. Thus, all electrical components, including extension cords, should be kept a safe distance off the floor.

In summary, take all necessary safety precautions to protect yourself and others. Be aware of all electricity and water combinations. Follow all National Electric Codes.

Fish Tank Selection

In summary, your fish tank choice will be influenced by space, budget, available time for maintenance, and aquaculture goals (desired yield). There is a wide variety of choices; deciding upon a fish tank can seem bewildering at first, but if you keep focused on what you desire to achieve out of aquaponics, then the size issue is easily resolved.

1. A 250-gallon (1000 liters) or larger fish tank has been proven to create a more stable system. Larger volumes are especially better for beginners because they allow more room for error. With larger volumes, changes to water quality happen more slowly than they do in smaller systems.

2. To raise a fish to "plate size" a fish tank volume of at least 50 gallons is required.

3. The recommended stocking density is one pound of fish per five to seven gallons of tank water (0.5 kg per 20-26 liters).

4. Determine fish tank volume from the stocking density rule above (one pound fish per five to seven gallons of fish tank volume or 1 kg per 40-80 liters).

Constructing an Inexpensive Tilapia Tank

A tilapia tank is nothing more than an above ground container filled with water. Examples for backyard tilapia farming use are swimming pools, IBC totes, fiberglass hot tubs, and lined plywood troughs. Of course, there are more industrial tank options (e.g., commercial fish tanks, concrete walls, etc.).

Tilapias need 0.5 cubic foot of water, or 3.74 gallons, for every pound of their body weight at harvest. So, if you want to keep 144 pounds of fish in the same pond, you will need to have one that holds 72 cubic feet of water, or 538.56 gallons. At harvest, average weight of a tilapia is approximately 0.5 pounds (220 grams).

A lined plywood trough that is 4 feet wide and 8 feet long, with 2.25 feet of water depth, would fit the size needed in this example. Of course, you can expand this pond to any size that you want. And there is nothing wrong with having more water volume per pound of fish than what is needed. Yes, your pump and filter will have to do more work, but unless the tank is extremely larger than what is necessary, the extra pumping and filtering that needs to be done is negligible. To figure out how many cubic feet, or gallons, you will need, decide on how many pounds of fish that you want to harvest every six to nine months, depending on species of tilapia fingerlings that you select, and then divide that weight in half to get the cubic feet. Then multiply the cubic feet by 7.48 to get gallons. For example: if you want to have 144 pounds of tilapia in your tank at harvest time, you take 144 and divide it by two to get 72 cubic feet (144 / 2 = 72). Then, to get the gallons, you just have to take the cubic feet, and multiply by 7.48 to get 538.56 gallons (72 x 7.48 = 538.56).

Once you know how many cubic feet or gallons you need, it is simply a matter of finding the right container to hold that much water; with a little lip at the top, so that your tilapia don't swim over the edge. To get the cubic feet of a rectangular pond, you multiply the length, times the width, times the water depth. To get the cubic feet of a circular pond, multiply the radius times the radius (R^2), times 3.14 (pi), then multiply the result times the water depth.

Determine Your Tilapia Tank Construction

The simplest solution to making a tilapia tank, is to use an above ground swimming pool. A twelve-foot diameter pool filled with 24 inches of water holds 1,696.46 gallons and is very inexpensive. However, there are a few of drawbacks to using one of these small pools as a tank. First, as the vinyl ages it tends to get very brittle, and it will eventually split down the side. You can support the sides after the pool is filled by surrounding it with thin plywood, or similar flexible material, but this will only buy you a second season of use. The best thing to do is replace the pool (or liner) every 24 months, or sooner. The second potential drawback to using a swimming pool as a tilapia pond has to do with whether or not the fish are affected by the vinyl leaching chemicals into their water.

A lined plywood trough has many advantages over other forms of ponds for home tilapia farming: It is durable and will last a long time. It can be built to any size. It makes harvesting tilapia extremely easy. It is relatively inexpensive to build. And, if you want to expand into a small commercial tilapia farm, it can be easily made with readily available FDA approved materials.

Wood Framed Tank Construction

Duct tape is used on plywood joints.

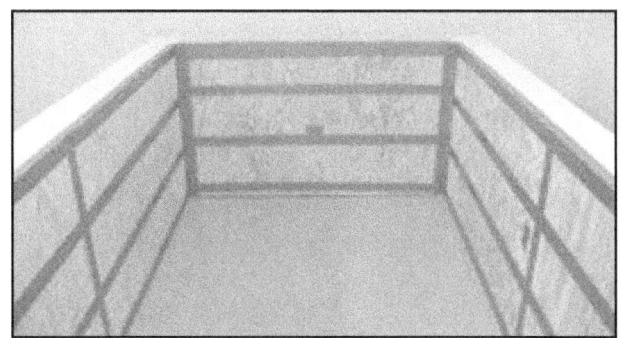

Finished wood tank with pond liner material for a watertight tilapia tank.

PART III : COMPONENTS USED IN AQUACULTURE

CHAPTER 8

Plumbing

Overview of Plumbing

Plumbing your aquaculture system requires careful consideration of many different factors and will depend on your own design and the type of aquaculture method being implemented. However, there are some basic principles that are applicable on almost every system which need to be applied. Therefore, issues pertaining to plumbing infrastructure are addressed in this chapter with the hope of providing you the best possible, user-friendly assistance.

Plumbing is an integral part of an aquaculture system and needs to be considered from the very beginning. In other words, one should consider plumbing to be just as important and as much as a priority as the fish tank. A common mistake is to develop the fish tank or pond, and then attempt to plumb it, afterwards only to find a major problem. This chapter, as well as sections in the following design chapters of this book, will help you avoid those time consuming, and sometimes costly, mistakes.

Water Conveyance

It should be understood that water pressure and water volume rate are two separate and distinct issues. Water pressure is the force that water flows from a plumbing fixture. Water volume rate is the amount of water present to fill a tank. Water pressure is typically noted in pounds per square inch (psi) or Kilopascal (kPa). Water volume rate is typically presented in gallons per minute (gpm), gallons per hour (gph), liters per second (L/s), or cubic meters per hour (m3h). Volume of water is normally described simply in terms of gallons, cubic feet, litters, or cubic meters.

Pipe diameter size will depend on system parameters, but it is always better to go with a slightly larger sized diameter pipe than the minimum size needed. There are a number of engineering related calculators online that can help you work out the ideal pipe diameter to use and it is prudent to review these during the planning phase of your system. The following are internet links to two such online calculators that can used to determine pipe size:

- http://irrigation.wsu.edu/Content/CalculatorsGeneral/Pipe-Velocity.php
- http://www.calctool.org/CALC/eng/civil/hazen-williams_g

Flow rate decreases as the length of pipe increases, so keep this in mind if conveying water or wastewater a considerable distance (e.g., an outdoor pond to a greenhouse is usually a considerable distance away, etc.). The line itself provides resistance to the water flow. Therefore, the length of the run is a major factor. The longer the run, the less gallons per minute that can flow through the line.

Length of run actually has a dramatic effect on the conveyance of fluid. As an example, a typical water line will lose approximately 33 percent of its water delivery capability when the length of the run is increased from 30 linear feet to 60 linear feet. As a specific example, 1-¼ inch diameter pipe can deliver approximately 21 gallons per minute over a run of 30 linear feet, yet only 14 gallons, approximately, per minute over a run of 60 linear feet. While the length of run is a major factor for water line size calculations, it becomes more of factor when the run is unusually long.

Table 17 is a helpful guide in determining the pipe size needed based upon the required amount of water that will be conveyed to empty the fish tank once per hour for filtration purposes.

Additional Pipe Sizing Considerations

Increasing the pipe to just one size larger makes a dramatic difference. Those not in the plumbing trade or engineering field do not realize that there is strong correlation between length and pipe diameter.

As an example, a 1-¼ inch diameter pipe is only 25 percent larger in diameter than 1-inch diameter pipe, but there is an area difference of 56 percent between these two slightly different pipe diameter sizes. Another example of this size/area relationship can be seen when examining the difference of the areas inside a 1-¼ inch diameter pipe compared to a 2-inch diameter pipe, which is about 77 percent.

When examining the flowrate (gallons per minute), the differences are even more dramatic. Basing calculations of an average run of pipe of 50-feet in length, a 1-¼ inch diameter pipe will convey up to about 16 gallons per minute. On the other hand, 1-inch diameter pipe only provides about 9 gallons per minute. Therefore, a 1-¼ inch diameter pipe provides

TABLE 17. **Water Flow Rate**

PIPE LENGTH (ft)	WATER FLOW RATE IN GLM Pipe Diameter in Inches									
	0.5	0.75	1	1.5	2	2.5	3	4	5	6
5	23	66	140	407	868	1560	2520	5371	9659	15601
10	16	45	96	280	597	1073	1733	3694	6643	10730
15	13	36	77	225	479	862	1393	2968	5337	8620
20	11	31	66	193	410	738	1192	2541	4569	7380
40	7	21	46	132	282	508	820	1747	3142	5076
100	4	13	28	81	172	309	500	1065	1916	3095

PIPE LENGTH (m)	WATER FLOW RATE, m³/hr Pipe Diameter in mm									
	12	20	25	40	50	65	75	100	130	150
1	5.6	21.5	38.6	133.0	239.2	477.0	694.9	1481	2953	4302
2	3.9	14.8	26.6	91.5	164.5	328.1	478.0	1019	2031	2959
4	2.7	10.2	18.3	62.9	113.2	225.6	328.7	700.5	1396.7	2034.9
6	2.1	8.2	14.7	50.6	90.9	181.3	264.1	562.8	1122	1635
12	1.5	5.6	10.1	34.8	62.5	124.7	181.6	387.1	771.7	1124
30	0.9	3.4	6.2	21.2	38.1	76.0	110.7	236.0	470.5	685.5

almost 77 percent more gallons per minute than a 1-inch diameter pipe.

What does all this mean to the operator? It means that for a nominal amount of money, increasing pipe diameter by just one size provides dramatic benefits. The table 18 clearly illustrates this point.

This is probably a good place to mention that the above noted plumbing principles apply to both tubing and pipes. There are minor head loss difference for each type of material being considered (e.g., PVC, steel, polyvinyl, etc.). Furthermore, fitting and bends also impact conveyance of the fluid being delivered. However, for the vast majority of systems, the head loss differences between material type and the number of fittings used are of such minor consequence that these differences can be ignored. Head loss needs to be considered, but not to the extent of examining each minor difference when planning and designing a system. Pipe diameter, elevation difference, and length of pipe run are going to be the critical factors that need to be considered in the planning and design phases.

Over time, debris can build up on the inside of the pipes and this will negatively impact the flow rate. Pipes may need to be cleaned periodically in order to ensure an unimpeded flow rate. The necessity to clean some sections of plumbing should influence your decision as to how you connect pipe and fittings (e.g., glue, installing a cleanout fitting(s) at certain plumbing points, etc.). Feeding a domestic garden hose through the pipe (with the water turned on) will sufficiently clean the pipe in most cases. A power washer does wonders, as well.

PVC Piping

The most commonly used plumbing materials used in aquaculture are PVC or UPVC pipes and irrigation tubing. Garden hoses are also used fairly often, especially when the fish tank is located a significant distance from the filtration unit.

Occasionally, there is an article questioning the safety of PVC, but after years of study by scientists all over the world, there is still little evidence to prove that it is harmful to fish or that it introduces toxins to the system. As such, it has been certified as safe for use in drinking water infrastructure by government agencies throughout the world. In reality, all evidence suggests that the risk of harm from using PVC pipes in food and water supply is so slight as to be negligible. Because of this, it is almost universally used for agriculture, aquaculture, hydroponics, aquaponics, and public potable water infrastructure.

Below are several beneficial features:
- PVC Pipe is almost universally available.
- PVC Pipe is usually extremely cost effective (low cost).
- PVC Pipe comes in standard sizes throughout the world.
- PVC Pipe has a wide range of adapters and connectors available.
- PVC Pipe is easy to use, cut and adapt.
- PVC Pipe is durable and long lasting.
- PVC Pipe is light weight.

Although other piping can be used in aquaculture (such as agricultural pipe, flexi-pipe, bamboo, hosepipe, etc.,) it important to make sure that it is safe for use in a system that grows food for human

Pipe Size (Sch. 40)	I.D. (range)	O.D.	GPM (w/ min. PSI loss & noise)	GPH (w/ min. PSI loss & noise)
1/2"	0.5 - 0.6"	0.85"	7	420
3/4"	0.75 - 0.85"	1.06"	11	660
1"	1 - 1.03"	1.33"	16	960
1-1/4"	1.25 - 1.36"	1.67"	25	1,500
1-1/2"	1.5 - 1.6"	1.9"	35	2,100
2"	1.95 - 2.05"	2.38"	55	3,300
2-1/2"	2.35 - 2.45"	2.89"	80	4,800
3"	2.9 - 3.05"	3.5"	140	8,400
4"	3.85 - 3.95"	4.5"	240	14,400
5"	4.95" - 5.05"	5.563"	380	22,800
6"	5.85 - 5.95"	6.61"	550	33,000
8"	7.96"	8.625"	950	57,000

Assume Gravity to Low Pressure. About 6 f/s flow velocity, also suction side of pump

TABLE 18. **Typical flow rates for common pipe sizes used in aquaponics.**

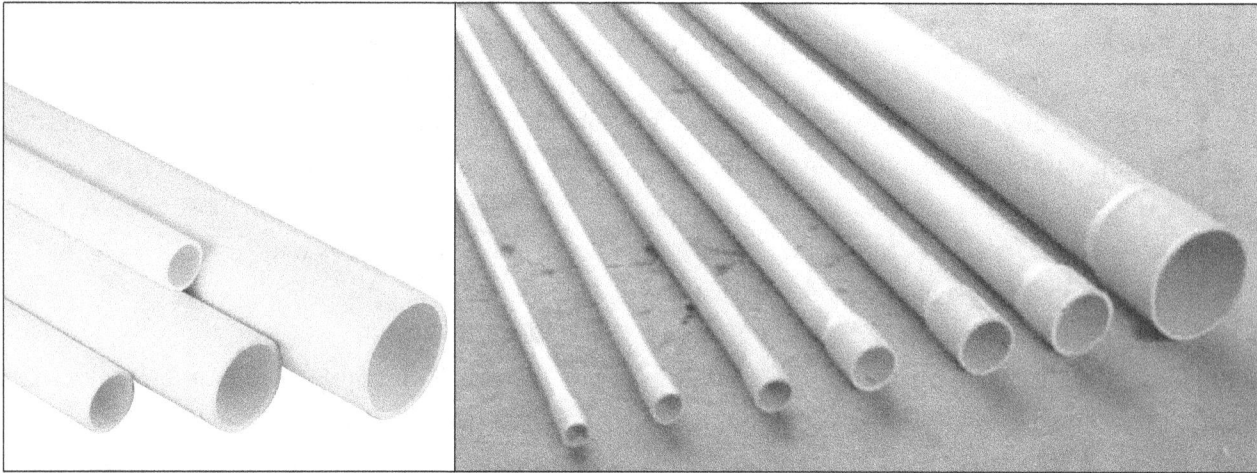

FIGURE 41. Straight (left) and bell end (right) PVC piping.

consumption, and make sure that it will not be harmful to fish. As an example, it is prudent to avoid using metal piping, especially copper piping, as it can be highly toxic to fish. Also, if installing used plumbing materials, make sure that it was not used to covey any toxic or harmful substances in the past. If in doubt, it is best to pass on it, especially since PVC pipe and fittings are relatively inexpensive.

Schedule 40 vs Schedule 80 PVC

If you've been shopping around for PVC you may have heard the term "schedule". Despite its deceiving title, schedule doesn't have anything to do with time. A PVC pipe's schedule has to do with the thickness of its walls. Maybe you've seen that schedule 80 pipe is slightly more costly than schedule 40.

Though the outside diameter of a schedule 80 pipe and a schedule 40 pipe are the same, a schedule 80 pipe has thicker walls. This standard of measuring pipe came from a need to have a universal system for referring to PVC. Since different wall thicknesses is beneficial in different situations, the ASTM (American Society for Testing and Materials) came up with the schedule 40 and 80 system for classifying the two common types.

The main differences between Schedule 40 (Sch 40) and Schedule 80 (Sch 80) are:
- Water Pressure Rating
- Sizing & Diameter (Wall Thickness)
- Color
- Application & Use

Water Pressure for Sch 40 vs. Sch 80

Both schedule 40 and 80 PVC are used widely around the world. Each one has its benefits in different applications. Schedule 40 pipe has thinner walls, so it is best for applications involving relatively low water pressure. In the vast majority of situations, schedule 40 is sufficient for aquaculture, aquaponics, and hydroponics.

Schedule 80 pipe has thicker walls and is able to withstand higher PSI (pounds per square inch). This makes it ideal for industrial and chemical applications. To give you an idea of the size difference, 1" schedule 40 PVC pipe has a .133" minimum wall and 450 PSI, while schedule 80 has a .179" minimum wall and 630 PSI. By comparison, the water pressure in most residential and commercial settings, as well as in aquaculture, aquaponics, and hydroponics rarely ever exceeds 125 psi, whereas, 30 to 80 psi is common in these applications.

Sizing & Diameter

As mentioned earlier, both schedule 80 and schedule 40 PVC pipe have the exact same outside diameter. This is possible because schedule 80's extra wall thickness is on the inside of the pipe. This means schedule 80 pipe will have a slightly more restricted flow, even though it may be the same pipe diameter as an equivalent schedule 40 pipe. This means schedule 40 and 80 pipe do fit together and can be used together if necessary.

The only thing to be careful of is that the lower pressure handling schedule 40 pipe meets the pressure requirements of your application. A pipeline is only as strong as its weakest part or joint, so even one segment of schedule 40 pipe used where a higher-pressure schedule pipe is needed can cause problems. However, it is rare to have such high pressures in aquaculture, aquaponics, hydroponics, and other agricultural situations.

Schedule 40 and Schedule 80 Color

Generally, schedule 40 pipe is white in color, while schedule 80 is often gray in order to distinguish it from 40. PVC is available in many colors though, so be sure to check labels when purchasing.

Which Schedule PVC do I Need?

So, what schedule PVC do you need? For aquaculture, home repairs, or irrigation projects, schedule 40 PVC is the way to go. Schedule 40 PVC is capable of handling impressive pressure, and it is likely more than adequate for light to moderately heavy applications.

Using schedule 40 will save you money, especially if you plan on using large diameter parts. For large commercial, industrial, or chemical applications, it would be wise to use schedule 80 pipe and fittings. These are applications that will likely cause higher pressure and stress on the material, so thicker walls are imperative.

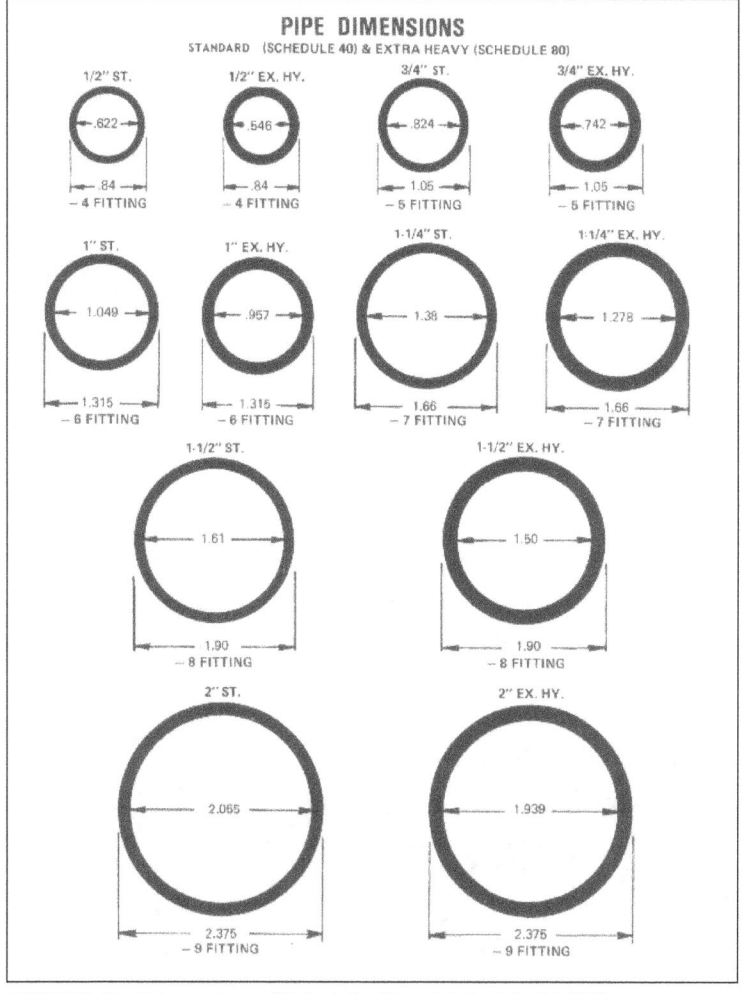

FIGURE 42. Dimensions of Schedule 40 and Schedule 80 PVC pipe.

Pipe Connections and Fittings

There are two commonly used PVC fittings in aquaculture, aquaponics, hydroponics, and other agriculture applications—threaded connectors and slip connectors. Threaded connectors are ones that screw into one another and are designated as male and female. Slip connectors, as the name suggests, just slip into one another. In order to preserve pipe diameter, avoid restricting flow, and minimize clog points, it is prudent to use only female fittings. Female fitting fit over the outside of the pipe, whereas male fittings fit inside the pipe. Examples of common connectors are:

- 90° elbows
- 45° elbows
- 90° Tee fittings
- Ball Valves
- Bulkheads
- Reducers
- Couplings
- Wyes

FIGURE 43. Standard PVC fittings and connectors

Irrigation Supplies

Irrigation materials are also used in aquaculture. There are several good reasons as to why irrigation materials are integrated into aquaculture. They are inexpensive, require minimal effort in learning how to install them, they are safe for fish and people, they are easy to install, and are effective conduits of aquaculture fluids.

However, not all irrigation materials are created equally. Some have thin walls and are so light they are basically a cheap low grade inferior solution. As inexpensive as irrigation materials are, it is better to invest in the higher-grade premium materials sold commercially to professionals from an irrigation supply store, rather than the typical residential stuff commonly sold at department and home improvement stores. Also, no more than what is typically used in a standard size system, the extra cost for premium materials is nominal. The result for paying slightly more for premium irrigation supplies is that your system will be much more reliable (less prone to leaks and not easily disrupted because of accidental damage). Premium irrigation materials are also much easier and faster to install than light weight residential materials.

It is also helpful to stick with the same name brand of irrigation materials, as they will be more interchangeable. Furthermore, it is helpful to shop at the same store so as to develop a solid working relationship with qualified staff. *Ewing Irrigation and Landscape Supply*, for instance, has expert staff on board eager to provide a wealth of information and assistance. And, no, I am not getting compensated for endorsing them. Nor do not have any friends or family working for *Ewing*, or benefiting from my praise of *Ewing Irrigation and Landscape Supply Co.* I say this as one consumer to another with the sole purpose of helping you. I have shopped for irrigation supplies for my nursery, aquaculture, hydroponics, and aquaponic operations for many years. I have been to many different supply stores (Ewing's competitors, several different large chain home improvement stores, Wal-Mart, etc.), and have tried many different products. *Ewing* won me over. Ewing has proven to me to have the best quality products, out of all the stores I have shopped, and they have provided me with the most knowledgeable professional advice. *Ewing* is located throughout the southern half of the United States, as well as the west coast. They also have a very user-friendly website and provide accredited education opportunities for those aspiring to learn more or development further professionally.

In summary, irrigation supplies offer a good alternative or supplement to PVC for plumbing of aquaculture, hydroponics and aquaponics systems. It is prudent to purchase premium quality irrigation products over cheaper residential type irrigation materials. It is also beneficial to shop at a store in which sound expert advice can be obtained and build a positive relationship with the staff. Following are some of the more common irrigation materials used in aquaponics:

FIGURE 44. 90-degree elbow, tee fitting, coupling

FIGURE 45. ¼-inch tube tee, ¼-inch tube elbow, ¼-inch tube coupling

FIGURE 46. **Tubing**, also called Poly Tubing, Poly Pipe, Supply Line, Trunk Line; all of which are common terms for this flexible polyethylene pipe (shown above). Common sizes are ½" (aka ⅝") or ¾" tubing. Emitters can be inserted into tubing or connected via ¼-inch diameter size micro tubing.

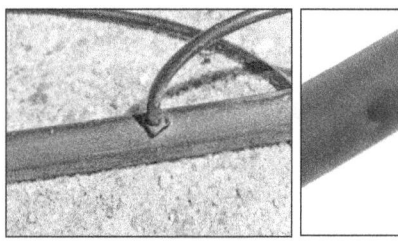

FIGURE 47. Two different types of ¼-inch tube to ⅝-inch tube connections

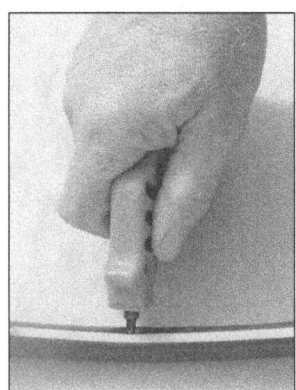 

FIGURE 48. Two different kinds of hole punches used to install ¼-inch tubing in larger dia. Irrigation poly tubing.

Hoses

In addition to piping and irrigation tubing, hosing can also be used for water conveyance in aquaculture, aquaponics, and hydroponics. The disadvantages of hoses are that they can kink, and some can be relatively expensive. Nevertheless, there is a wide variety of hose material types that can be used (see figures 49–53). Having clear (see through hose) will provide you the benefit of being able to watch your system at work, which is both helpful and fun. Use a hose clamp (figure 54) to obtain a secure watertight connection of the hose to pipes and fittings.

Fluid Dynamics

There are a few things to consider that influence how fluid actually flows through a pipe. The pipe and fittings cause friction to what would otherwise be smooth movement of the water, so the water in the very middle of the pipe is conveyed somewhat faster than the water flowing near the sides of the pipe. The difference is small, but it exists, nonetheless. In addition, comparing the flow rate through a straight length of PVC pipe, versus that of a pipe with a series of bends, the water flows more quickly through the straight pipe. There is no need to use complicated engineering equations; just recognize that there are some factors that will determine the flow rate through the plumbing system, which in turn impacts what size diameter piping or tubing should be used. Simply put, in a given amount of time you can move a greater volume of water through a big pipe, hose, or tube than you can through a smaller. Also, if moving water over a significant distance then it is especially important to caution on the side of using larger diameter piping or tubing instead of a smaller size, even if a reducer must be used at both ends.

Many pumps suitable for an aquaculture system come with a ½-inch outlet. Rather than convey the pump outflow through a ½-inch irrigation line, it is better to install a reducer at the pump. For example, at the pump, install a short ½-inch irrigation tube

FIGURE 49. Flex hose

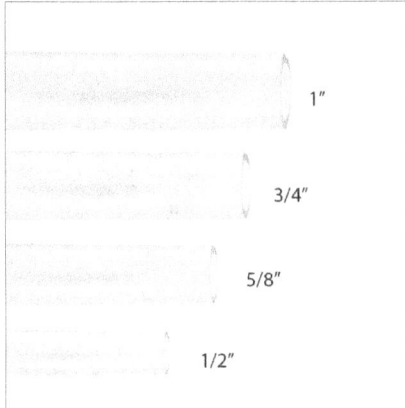
FIGURE 50. Clear vinyl tubing

FIGURE 51. Reinforced PVC hose

FIGURE 52. Braided PVC hose

FIGURE 53. 5/8 to 1-inch diameter garden hose

FIGURE 54. Hose clamp

about an inch in length, then add a ½-inch to ¾-inch reducer, and connect it to a ¾-inch poly irrigation tube to convey the fish tank wastewater to the filtration system and back to the fish tank.

Another thing to consider is gravity. Gravity will exert a constant downward pressure on the water in the system. A pump used to lift water will be a fight against gravity. This means that the amount of water a pump can push will be reduced the more it has to lift the water. This will be discussed in more detail in chapter 11, which addresses pumps.

Gravity can also be used to provide a great advantage. It can be used to move water with no mechanical intervention. For instance, if a filtration system is higher than the fish tank, then the overflow water can return to the fish tank directly through the use of gravity in a well-designed system.

When considering plumbing, think about all the water that will be in your system. Such includes the water in the fish tank, filtration system, and in the pipes. The sum of that volume is the amount of water in the aquaculture system. As a general rule, for a small to mid-size Tilapia tank, the entire volume of water in the fish tank(s) should be moved every hour in order to maintain good water quality for the fish. This issue will be addressed in more detail in the water quality, pump, and chapters.

As discussed in this chapter, and addressed in more detail later in the book, the 'head' of the system is how high in elevation the pump must move the water. Another item to consider is whether the system is to have a pump running continuously, or if it will have a timed flood and drain system. All of these factors together determine how much water needs to be moved, and the time it takes to move it. Therefore, due diligence is important in plumbing design and pump size selection. The pumps, pipes and/or tubing need to be large enough to sufficiently convey the required volume of water, but not oversized to the point where other problems are created.

Also, a well-designed system needs to have various controls and safety measures in place just in case there is any problem. Not to worry, as these issues can easily be determined, and they will be addressed in a user-friendly way later in this book. This chapter is meant to serve only as an introduction into these issues, and to emphasize that when selecting piping and/or tubing, it is better to pick slightly larger diameter plumbing component when possible.

Watertight Pipe Protrusion through Fish Tanks

To achieve a watertight connection when passing a pipe or polyvinyl tubing through a fish tank wall, a bulkhead fitting or a Uniseal can be used. Bulkheads are sturdier and work better for thick-walled applications, but also cost more than Uniseals.

The average price for a PVC bulkhead is approximately 25 percent more than a Uniseal.

Also, additional fittings are needed to connect the pipe to the bulkhead, whereas with a Uniseal, the pipe will slide through the opening (no additional fittings are needed). For additional protection or if a drip leak is discovered after installation, with a Uniseal or a bulkhead fitting, heavily applied silicon caulking will often resolve the problem.

Bulkheads

Bulkheads come in a very wide variety of sizes and shapes but can be easily assembled from parts readily available in most plumbing supply stores and home improvement centers, as well as online. Bulkhead fittings are identified by the size of the pipe it connects to, not the hole size. The bulkhead is a good, sturdy option for plumbing through a fish tank or grow bed. Bulkhead fittings are typically made of polyethylene, CPVC, PVC, polypropylene, Polytetrafluoroethylene (PTFE), and of various metallic materials. All types will work for aquaponics, so it makes sense to select the more com- mon and inexpensive types, such as polyethylene, CPVC, or PVC.

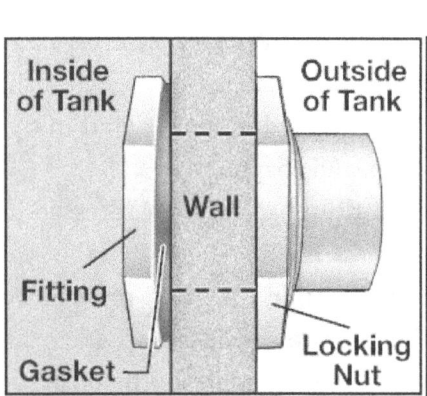

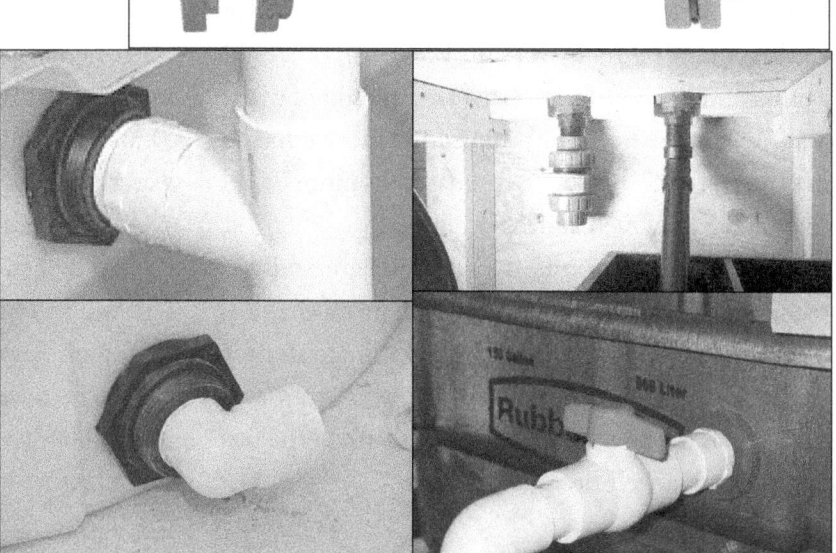

FIGURE 55. Bulkheads

Grommets (Uniseal®)

Grommets (Uniseal®) are rubber rings that fit into the holes that have been drilled into the tank. Uniseal is a company trade name. A uniseal is a grommet, but they are so common that most people refer to all grommets as 'unseals'. It is similar to people referring to drywall as sheetrock. Sheetrock is a company name. Drywall (also known as plasterboard, wallboard, sheetrock, gypsum board) is a panel made of gypsum plaster pressed between two thick sheets of paper. It is used to make interior walls and ceilings. Since most people in aquaculture and plumbing refer to a grommet as a 'unseal', the same will often be done throughout this book so as to hopefully help the reader better relate to the message being provided.

A uniseal clamps around the hole making a watertight connection; the PVC pipe can then be slotted into the seal. The seals usually allow the pipe to be installed in only one direction, thus providing a watertight seal between the pipe and the connector. Unseals are inexpensive, costing anywhere from about $2 to $15 (USA, year 2021), depending upon quality, size, and where it was purchased. They make it easy to put a pipe through a tank, and they can also be used with rounded surfaces thus making them particularly useful for plumbing into barrels and other like rounded containers.

Uniseals will even allow you to plumb directly into five-gallon buckets, brute trash cans, or any type of round surface. They accept standard Schedule 40 or 80 PVC and allow DIY projects that once may have been very costly to complete much more affordable.

Uniseals are used to attach pipe to just about any container in situation where bulkheads will not work, or not preferred. The most common use for uniseals is on curved surfaces such as storage drums, buckets, or even other pipes.

The advantage to using a uniseal is that it is inexpensive and provides for a quick and easy installation. A hole is cut in the side of the tank, and then the rubbery black uniseal is inserted. Next, a slippery

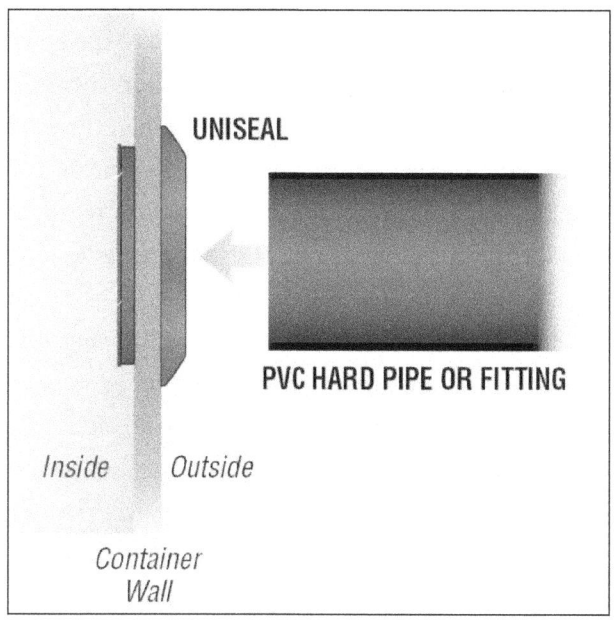

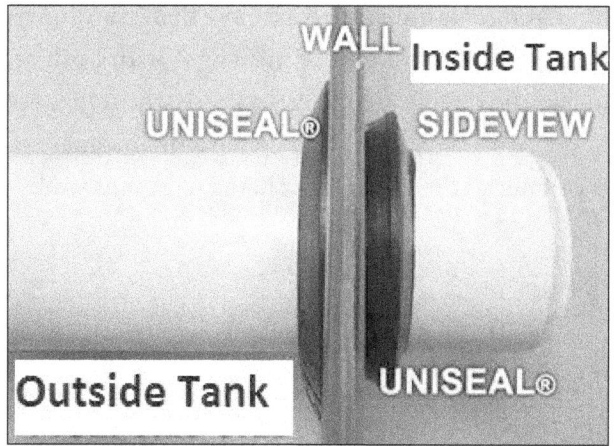

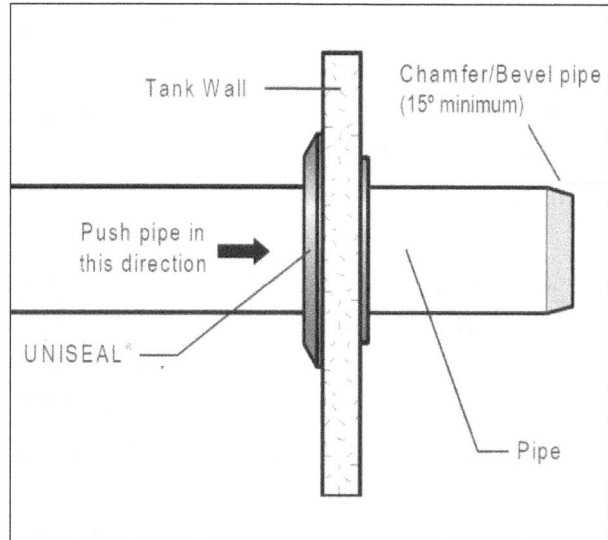

FIGURE 56. Grommets (Uniseal®)

detergent film (e.g., dish soap) is applied to the exterior of the pipe. The pipe is then pushed through the seal completing the installation. When the pipe is pushed through the rubber-like uniseal from the outside, the uniseal becomes thin enough to allow the pipe to slip through. With this simple design, the uniseal solves many complex problems.

The disadvantage to using a uniseal is that it is prudent to replace them anytime you need to take out or reinstall the pipe, as they lose their watertight structural integrity. Also, they cannot be used in thick-walled applications. Lastly, in high pressure situations (for instance, in extremely tall tanks) a bulkhead would be a better choice.

Grommets (uniseals) have more than one dimension to consider. They are measured with both an inside and outside diameter; basically, the thickness of the "wall" changes. Plus, there is the thickness of the material the grommet is being used it as well.

Watertight Pipe Protrusion through a Liner

There are a number of devices that can be used to ensure a watertight seal where a pipe protrudes a liner. One item is a PVC liner boot. The PVC liner boot can either be built into the liner or made separate for field installation. A pipe coupler can also be used. A conduit flashing pipe boot (cone image below) also works well. The items can easily be obtained from your local plumbing supply store, hardware store, or home improvement center at a relatively low cost.

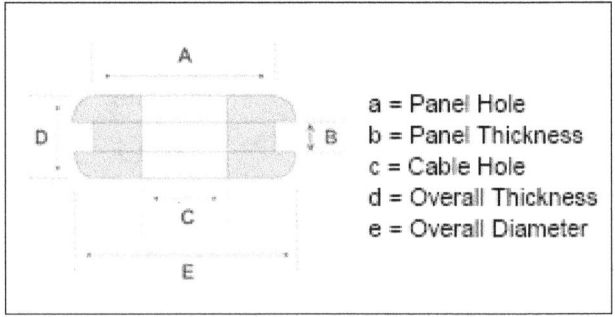

FIGURE 57. *Grommet (Uniseal)*

Uniseal Specifications

½" Uniseal
Fits Schedule 40 or 80, ½" pipe
Pipe ID — ½"
Hole saw size: 1 ¼" or 31.7mm (32mm)

¾" Uniseal
Fits schedule 40 or 80, ¾" pipe
Pipe ID — ¾"
Hole saw size: 1 ¼" or 31.7mm (32mm)

1" Uniseal
Fits schedule 40 or 80, 1" pipe
Pipe ID — 1"

Hole saw size: 1 ¾" or 44mm

1¼" Uniseal
Fits schedule 40 or 80, 1¼" pipe
Pipe ID — 1¼"
Hole saw size: 2" or 50.8mm

1½" Uniseal
Fits schedule 40 or 80, 1½" pipe
Pipe ID — 1 ½"
Hole saw size: 2½" or 64mm

2" Uniseal
Fits schedule 40 or 80, 2" pipe
Pipe ID — 2"

Hole saw size: 3" or 76mm

3" Uniseal
Pipe ID — 3"
Hole saw size: 4" or 102mm

4" Uniseal
Pipe ID — 4"
Hole saw size: 5" or 127mm

6" Uniseal
Pipe ID — 6"
Hole saw size: 7" or 177.8mm

Uniseal Installation Instructions

- Cut hole to the hole saw size indicated for the uniseal below.
- Ensure that the hole is clean with no sharp edges. Irregularities can cause a poor seal and leaks.
- Insert the uniseal into the hole with the wide flange on the outside of the container.
- Ensure the pipe end that will be inserted is clean of burs or sharp points. File the edges if needed.
- Insert the pipe into the uniseal. You can lubricate the pipe end with Windex. Most of it will be squeezed off during installation, but be sure to wash it out thoroughly after the pipe is inserted.

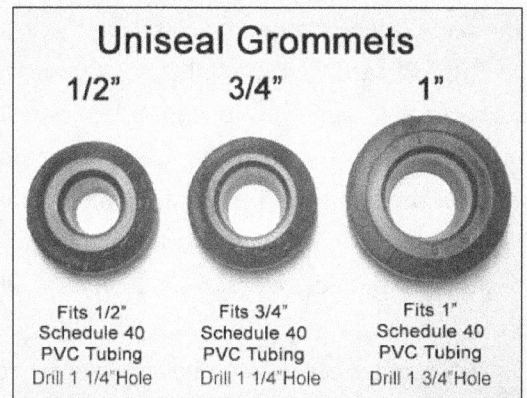

FIGURE 58.

Pipe Maintenance

Periodic cleaning of pipes is a maintenance chore that should not be neglected. It will ensure that your system continues to run efficiently, and that no unnecessary stress being placed upon pumps, and fish. Pipe cleaning will remove blockage due to accumulation of algae, mosses, or other debris. Since each system is different, there is no hard and fast rule as to how often this should be done, but it is prudent to do it sooner and regularly rather than waiting until a problem occurs.

A drain snake is the most standard pipe cleaning method used. Drain snakes are available as either a manual or power operated tool. Running water through the pipe during and after the process is recommended. However, it is best to detour the cleaned waste debris out of your system, or catch it in a bucket, rather than allowing it to enter back into your system. Never use chemicals.

A power pressure spraying machine is ideal for cleaning pipes. They can typically be rented from home improvement centers, as well as tool rental stores.

FIGURE 59. *Several methods for making a watertight pipe protrusion through a liner.*

A built-in debris trap located underneath the fish tank will greatly help reduce accumulation of pipe clogging debris. The following figure provides several examples of debris traps and cleanouts:

The fish tank is a crucial part of your system and so its size, safety, water quality, strength and long-term usability must be carefully considered. As such, the fish tank is typically one of the most expensive components of the system. However, there are some cost-effective options, which we'll address in this chapter.

Fish require certain conditions in order to survive and thrive, and therefore the fish tank should be chosen wisely. There are several important aspects to consider including the shape, material type, placement, and color.

If the fish tank is to be located outside, it will be subject to environmental conditions. It needs to withstand direct sunlight (UV resistant) and temperature variations without cracking, warping, or leaching chemicals into the water. For best results, shade fish tanks to prevent algae growth and to reduce stress to the fish; they prefer dark hiding places. Avoid direct lighting on the fish tank.

If possible, in addition to shading the top, place at least one object within the water that will provide the fish a sense of security. Such is rarely done in medium to large scale aquaculture operations. However, an object in the water, which offers fish refuge, will go a long way in improving fish health via reduced stress, and is considered by many to be

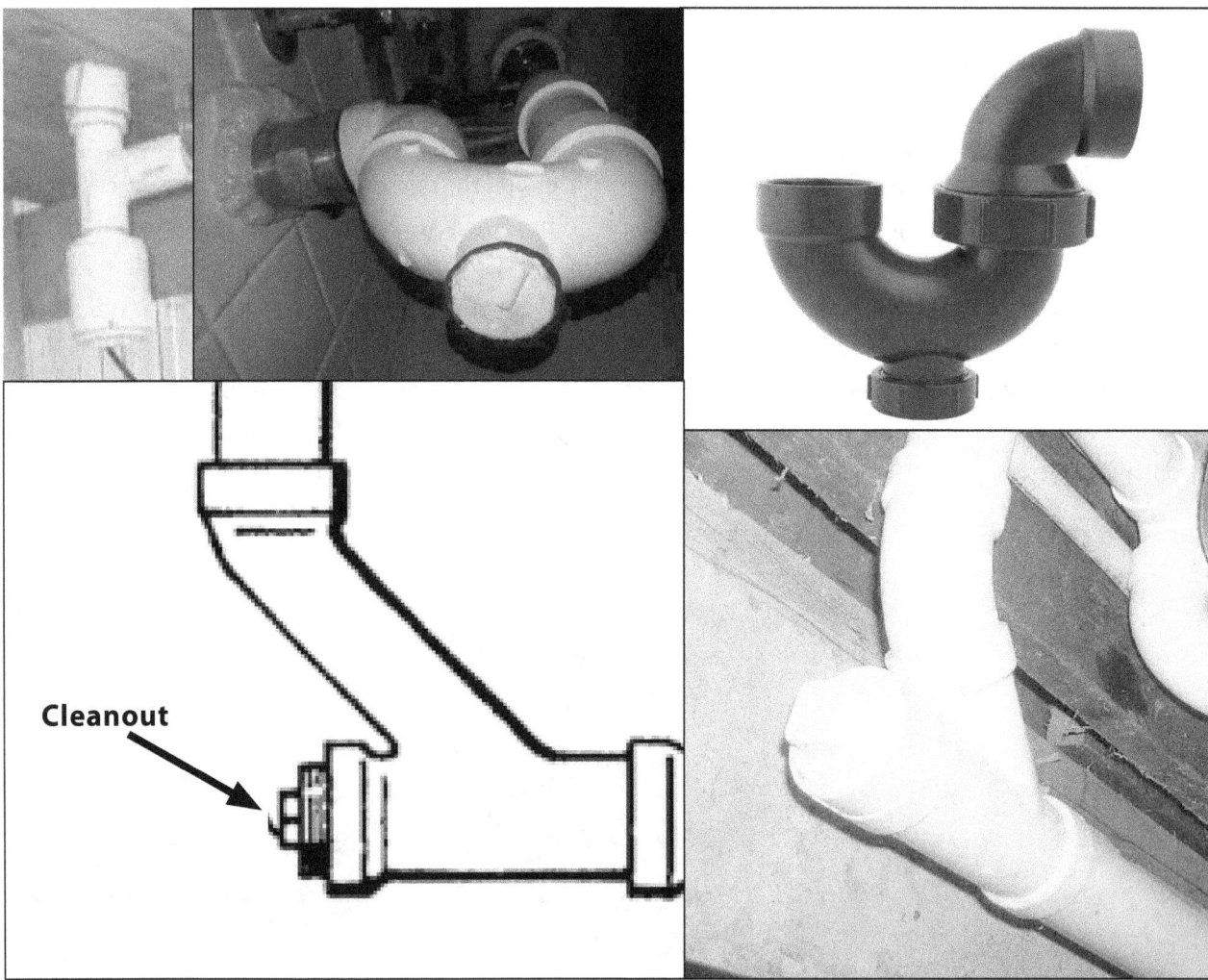

FIGURE 60. Debris traps

a more natural, animal friendly, and ethical means to rear fish, despite the size of the tank or operation.

Your choice of fish tank will also depend largely on your goal of aquaculture. If you are building a small size aquaculture system for hobby purposes, then you will be restricted to growing fewer fish. If you desire to rear fish for food, the fish tank should be large and sturdy enough to hold at least 50 gallons of water. The fish tank should be made of food-safe materials. However, fish tanks can be constructed from just about any structure or recycled materials, so long as they are lined with LDPE, PVC, HDPE pond liners, or other safe materials. Do not use EPDM liners or products in your system as EPDM releases toxic chemicals over time that are harmful to the living organisms within your system and can also enter into your food supply.

As a general rule, use approximately one pound fish per six gallons of water, for fish stocking densities. This will determine the volume of water required for the size of the fish tank. Higher stocking densities per volume will cause the accumulation of toxic waste, and result in the loss of fish as there would not be enough bacteria and plants to remove the nutrients fast enough.

PART III : COMPONENTS USED IN AQUACULTURE

CHAPTER 9

Liner Material

Liner Overview

A liner is an impermeable geotextile material used for water retention. Liners are commonly referred to as pond liner, water garden liner, greenhouse cover material, hydroponics pond liner, aquaponics bed liner, and polyethylene tarp material. However, liners are also used in many other types of industries in a wide variety of applications.

Liners are often used in aquaponics and aquaculture for flood-and-drain grow beds, D.W.C. tanks. fish tanks, and fish ponds. With the aid of structural support, a liner has the ability to contain large volumes water, even against significant pressure. The advantage of using a liner over other materials is lower cost, ready availability, and ease of use.

There are many different types of liners available on the market. As matter of fact, there are well over 24 different types of common liner materials. However, only some materials are suitable for aquaculture.

Acceptable materials include LDPE (Low Density Polyethylene), PVC (Poly Vinyl Chloride), PVC with internal reinforcement, and HDPE (High Density Polyethylene). HDPE is frequently noted as the best, but it is the most expensive and hardest to work with because of its ridged nature. LDPE is most preferred by small to mid-size aquaculture operators.

Avoid EPDM as it emits toxins that are harmful to the beneficial bacteria in your system and the toxins will eventually make their way into your food supply. Also avoid vinyl liners as they are too stretchy as well as any product that doesn't specifically identify what is made of.

The liner thickness should range anywhere from 20 to 40 mils. Obviously, the thicker the better in terms of protection against tears. However, the thicker the liner, the more difficult it is to handle and work with during installation, and more expensive it will be.

The ideal liner will be UV resistant, fiber reinforced, have one side white for easier installation, be of food grade quality, and be of thick enough to avoid tears during installation or regular use.

Liner Installation Precautions

Whether installing a liner in a pond, tank, or grow bed, care needs to be taken to ensure that the piece is large enough to do the job. Allow excess to overlap grow bed rims and tanks or to be buried in the ground for anchoring purposes. In other words, measure twice, cut once.

Refrain from walking on or laying tools or supplies on the liner. Avoid installing liner material outdoors on windy days or make sure you have enough people to assist you. Also, don't let pets walk on the liner material either. Any of the above can cause a puncture and result in a major hassle when discovered after the system is filled with water.

Liner Modifications

Occasionally, two separate pieces of liner material need to be joined in order to have a piece large enough to do the job. The following are directions from the Colorado International Lining Company (www.coloradolining.com) for field seaming liner material.

RECOMMENDED GUIDELINES FOR FIELD SEAMING

Temperatures for seaming should be as follows:
- Minimum ambient temperature: 50° F.
- Liner surface temperature: 90-100° F.
- Solvent adhesive should be 70-80° F.

1. Before adhesive is applied, surfaces to be seamed must be dry and free of dirt and foreign materials. The contact surfaces of the panels should be wiped clean to remove all dirt, dust, or other foreign materials. A clean cloth or grout brush may be used to help clean.
2. Adhesion of one liner panel to another is accomplished by lapping the edges of panels a minimum of 3 -6" inches (6" up to 1' is more desirable).
3. If using a seaming board, it should be placed below the liner panels to be joined. Only the pull rope should be exposed through the seam.
4. To commence seaming, fold the upper sheet back and apply adhesive to both sheets at the overlap area using a 3" wide paintbrush. Only apply about a body width (shoulder to shoulder width) of adhesive at a time. The adhesive should be applied to reach the outer edge of the seam and be 2-3" wide (within the overlapped area). Use enough solvent to make the liner surface appear wet and shiny (not flooded or inundated). Once the solvent is brushed-on, the brush should be replaced in a can (not on the liner) and the seam be rolled flat with a seam-roller once both surfaces become tacky. Roller strokes perpendicular to the liner seam should be made to push out excess adhesive and air pockets. Some adhesive should barely start to ooze from under the top overlapped edge. Once 100 percent is burnished down, the next shoulder to shoulder width area can be adhered with the same repeated process. The seaming board shall be advanced as needed.
5. Always keep the solvent adhesive can sealed. Only remove the lid long enough to wet the seaming brush. Failure to do so will result in the adhesive losing its "solvent grab" through evaporation of volatiles and seam quality will deteriorate to unacceptable quality. One option is to have a separate paint style can with a slit carefully cut into the lid allowing the brush to fit snugly within. Once brushes begin to stiffen-up, they should be replaced.
6. Using too much PVC bonding solvent in a wide area could result in a poor quality seam and make repairs difficult.
7. About one hour after seaming is performed it is advisable to manually probe the seam. Any loose edges should be peeled back and re-glued or patched if necessary. All repaired areas should be identified with a crayon or marker for re-inspection after glue dries.
8. Seams typically require 24 hours to completely cure. They can reach 90 percent of their final cure strength in just a few hours.

CHAPTER 9: LINER MATERIAL

TOOLS SUPPLIES LIST FOR FIELD SEAMING

- PVC bonding solvent specially formulated for liners (i.e. LO-VOC X-15™ PVC Solvent, or equal). Purchase enough for your job. Read product label for coverage specifications.
- Typical solvent adhesive yield is: One gallon of adhesive should bond 75' — 100 lf of seam.
- Safety glasses or safety goggles.
- Rubber gloves.
- Well ventilated area.
- Knee pads.
- Cotton rags.
- Several 3" wide paint brushes
- One (1) whisk broom (or fox tail brush).
- One (1) 10" x 6' -8' long Douglas Fir clear board, rounded off on both ends as well as all edges, with a rope tied to one end.
- One (1) Stanley Knife.
- One (1) yellow crayon for marking liner surface.
- One (1) pair of scissors with rounded-off points.
- One (1) 2' plastic or wood wallpaper seam-roller per each field seaming crew.

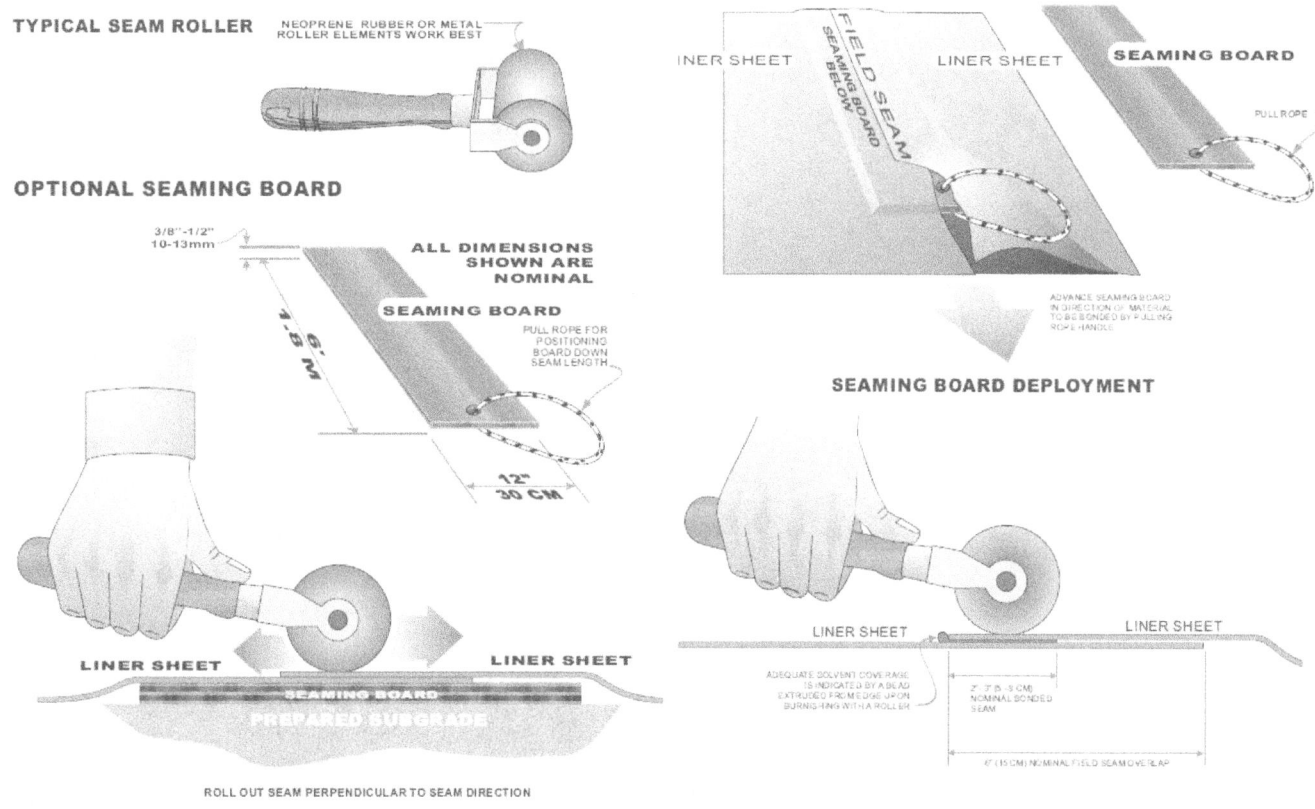

FIGURE 61. *Joining two liners with a watertight seam.*

Liner Repairs

To repair small rips or puncture holes, follow the instructions for sealing a seam above, but using the method described in the illustration (see figure 62).

Liner Plumbing Penetrations

Although this topic is addressed in greater detail within 'Chapter 8: Plumbing', along with other watertight pipe penetrations, figure 63 shows a watertight pipe penetration through a liner.

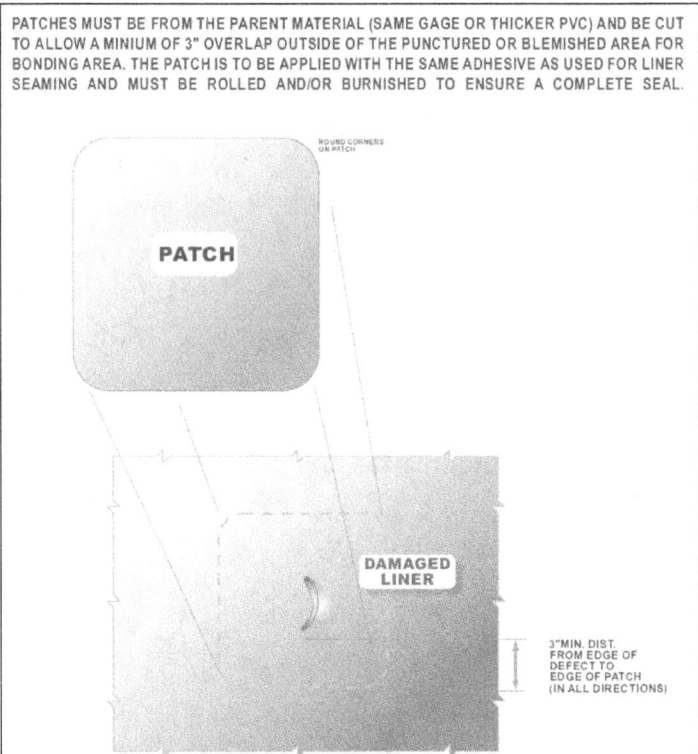

FIGURE 62. *Patching a damaged liner.*

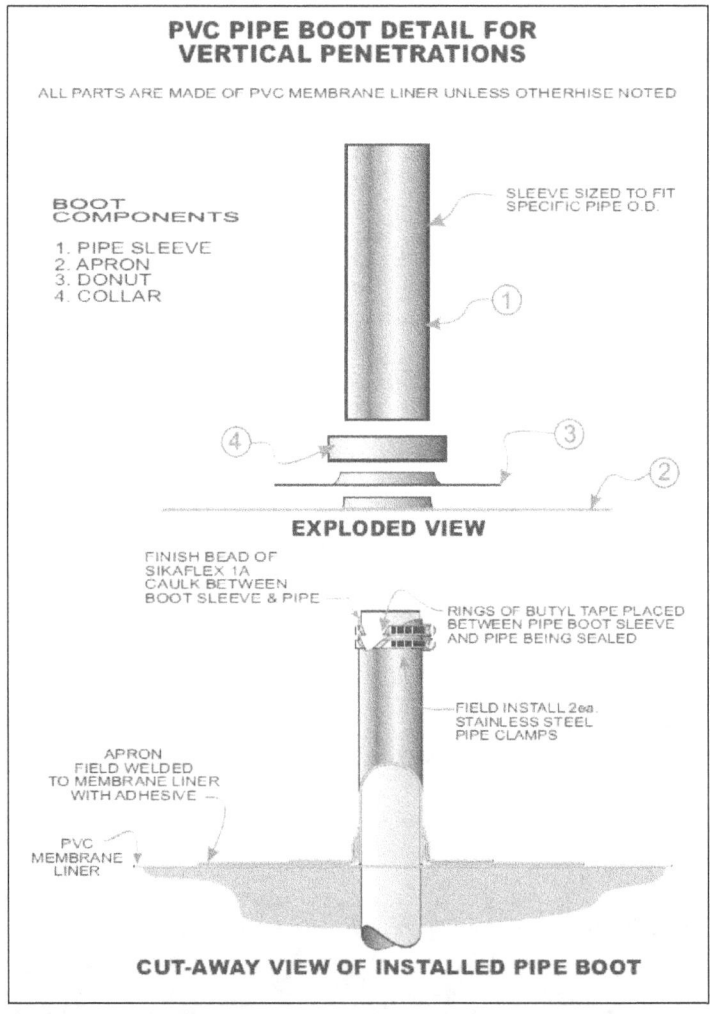

FIGURE 63. *Making a watertight pipe protrusion through a liner.*

CHAPTER 10

Making a Water Tight Container

Making a Watertight Wood Frame Container (DIY)

In addition to using a liner, as discussed previously, there are a variety of other waterproofing options available. The waterproofing materials discussed in this section are in the epoxy or resin categories and adheres to the wood structure. There are a variety of options available to seal a wood framed tank. Structural support is primarily achieved via a wood frame. Plywood and/or other boards are typically used for the walls and floor. The next step is to waterproof the container so that it will hold water.

The point of this section is not to endorse a particular product or method, but rather to provide comprehensive, unbiased, and reliable information about the primary methods that can be successfully implemented so that you can ultimately make your own educated decision. The following are other main methods in which to make a water-tight, sealed container (e.g., fish tank).

1. Two Part Epoxy Paint

A commonly used and reliable epoxy paint for plywood container builds is 'Sweetwater' epoxy available from Aquatic Ecosystems. It is non-toxic when cured and has good adhesion to lumber materials. This epoxy paint has solvents mixed in and is only 65-72 percent solids. It has strong, toxic fumes while curing, so a well-ventilated workspace is essential when using this product. It has a nice consistency and is quite easy to work with and apply. Sweetwater can be used as a stand-alone product for waterproofing a wood container. It is important to have a well-built structure when using this product as excessive flex could potentially contribute to leaks. Silicone will adhere well to Sweetwater so when sealing a tank with this method you should first apply the Sweetwater and then silicone (e.g., fish viewing windows, grommets, bulkheads, pipes, etc.).

Epoxy Paint Summary

PROS: Easy to apply, available in a range of colors, silicone will stick to it making window installation easy.
CONS: Toxic solvent requires well ventilated workspace, does not add structural strength to a container. Potential risk of failure in an inadequately supported tank with excessive flex.

2. Fiberglass Resin with Epoxy Paint Top Coat

Reinforcing a wood container with fiberglass is an excellent way to add structural strength to the build and impact resistance to a non-structural epoxy coating. An inexpensive way to apply fiberglass to a wood container is to apply a fiberglass resin to wet out the cloth. Lightweight fiberglass cloth and fiberglass resin are available from online vendors and are also generally easily found in home improvement centers such as *Home Depot* and *Lowes*, as well as in hardware stores such as *Ace Hardware*.

Fiberglass resin is generally a polyester resin and requires a small amount of hardener to be added as a catalyst. Effective application of fiberglass takes a little skill but is fairly easy to learn with some practice. It is important to avoid bubbles in the fiberglass layer, which may ultimately pop and compromise the watertight seal. Polyester fiberglass resin is fairly inexpensive. It has an extraordinarily strong toxic smell and requires a well-ventilated workspace. In addition, it is not sufficient as a standalone waterproof barrier coating and will leach out chemicals into your system. Therefore, it is important to finish a fiberglass resin coated container with a non-toxic, waterproof topcoat. *Sweetwater* epoxy is an excellent product for this purpose. Using *Sweetwater* over a layer of fiberglass overcomes most of the potential cons of just using *Sweetwater* alone.

Fiberglass Resin + Epoxy Paint Summary
PROS: A relatively inexpensive way to add structural strength and impact resistance to a wood container. The epoxy topcoat is available in a range of colors. Silicone will adhere to the epoxy topcoat.
CONS: Fiberglass application requires some practice and skill. Toxic fumes in the resin and paint require well ventilatedworkspace. Fiberglass resin is not sufficient as a stand-alone barrier coating.

3. Two-Part Marine Epoxy Resin

There are many different brands of epoxy resin available, but only true two-part marine epoxies should be used for wood frame aquaponic containers. These epoxies have a long established and successful history in waterproofing wooden boats. They differ from epoxy paints in that they are 100 percent solid. Several brands are available, but some tried and true options include:
- *West Systems 105* (one of the more expensive products).
- *US Composites* (an inexpensive option).
- *Max ACR* (relatively affordable epoxy being marketed specifically for aquatic applications)

There are many other marine epoxy brands available that can be used in aquaponics and aquaculture. The three noted above cover the spectrum of price and have all been successfully used in waterproofing wood containers.

Marine epoxies come as two-parts, a resin, and a separate hardener, which have to be mixed in a precise ratio. It is best to use slow hardeners when sealing a wood container with these products so as to be provided a longer working time and better penetration into the wood. It is strongly recommended that instructions be thoroughly read, and a 'how to' video on the internet be watched before beginning to work with these products. An instructional video will be posted on the **www. FarmYourSpace.com** website.

Marine epoxies can be used as a standalone product to provide a completely waterproof and non-toxic coating for wood containers. However, they are fairly brittle when cured, can be susceptible to stress fractures at seams, and dam- age from impact which will compromise the barrier coating. The best way to avoid these issues is to incorporate a layer of fiberglass cloth into the epoxy resin. Epoxy resins can be used to wet out fiberglass cloth in much the same way as polyester fiberglass resin but offer several advantages.

There is not a strong smell. The cured resin layer is completely waterproof, non-toxic, and is slightly less brittle than polyester resin. With the exception of cost, epoxy resin is an all-around better option than polyester fiberglass resin. As discussed above, wetting out fiberglass cloth does take a little practice, but with a little skill is an excellent way to add significant structural strength to a wood container.

When used with thickening agents such as colloidal silica, marine epoxy can also function as an excellent adhesive, particularly for slightly loose joints or joints where high clamping pressure cannot be applied. It can therefore also be useful as a waterproof adhesive during the construction and assembly of a wood container.

Silicone will adhere well to epoxy resin, so waterproof the wood container first and then silicone in your any viewing windows, bulkheads, grommets, etc. Epoxy resins can be tinted with a coloring coating agent but will usually still show wood surface grain and joints. Sweetwater epoxy, mentioned above, can be used as a topcoat if solid color is desired.

Two-Part Marine Epoxy Summary
PROS: Can add significant structural strength to a wood container, particularly when used with fiberglass. Minimal smell. Effective standalone waterproof barrier layer. Silicone will stick to it making window installation easy.
CONS: Expensive and it can suffer from stress fractures if used without fiberglass on an inadequately supported structure.

4. Pond Shield

Pond Shield is a 100 percent solid, two-part epoxy resin available from Pond Armor. It is different enough from the marine epoxies described above that it is discussed here as a standalone waterproofing option. *Pond Shield* is non-toxic, is almost odorless, and is safe to apply indoors. *Pond Shield* is an extremely thick epoxy which can make it somewhat challenging to work with. The black pond shield is the thickest and has a consistency slightly thinner than honey. It can be thinned marginally with denatured alcohol to make it more workable but thinning also increases the risk of not getting the required thickness in a single coat. One 1.5qt kit of *Pond Shield* covers 60sq ft at 10mil thickness. For many applications, a single coat is sufficient to get a 10mil coat that is completely waterproof.

Pond Shield is best suited for waterproofing well-supported structures with no flex. If there are concerns about the structural integrity of seams, it is best to fiberglass them in order to prevent the formation of potential stress fractures. *Pond Shield* is ideal for sealing concrete and masonry tanks. While it adheres to wood, it holds even better to concrete and masonry. Therefore, one option when using it on a wood tank is to first line the inside with *Hardiboard*. If you instead choose to apply it directly to a wood tank, it is best to first do a light wash with 30-40 percent alcohol-thinned *Pond Shield* to get better penetration into the wood. Follow this with a single coat of un-thinned, or very slightly thinned, *Pond Shield*.

The thick consistency of *Pond Shield* can make it challenging to work with and may require touching up in some areas after the initial coat. Careful inspection and touch up is critical to the success of using this product.

Pond Shield Summary
PROS: Completely non-toxic and odor free. Available in a range of colors. Requires only a single coat and touch up. Silicone will stick to it making window installation easy.
CONS: Thick consistency can make it difficult to work with. While it remains somewhat flexible after curing, it is susceptible to stress fractures in poorly supported structures.

5. Liquid Rubber

These waterproofing products are elastomeric emulsions that remain highly flexible after curing. While there are a number of these products available, some examples of those that have been successfully used with wood containers are:

- *Zavlar* (named to Permadri Pond Coat in the US).
- *Ames Blue Max*.

These products have minimal odor and are easy to apply. *Zavlar* (Permadri Pond Coat) requires several coats to achieve a waterproof barrier of 40mil thickness. One gallon will cover 30 square feet at 40mil thickness.

The primary advantage of these products is that they are incredibly flexible, which allows them to stretch and resist fractures in inadequately supported structures. However, it is still a good idea to use drywall tape on seams while applying these products in order to reduce potential stress on the coating at these spots.

Zavlar (Permadri Pond Coat) will initially appear brown but will dry to a black coating. However, this black coating will gradually turn back to brown after being submerged for a period of time. Ames Blue Max dries to a translucent blue color.

Zavlar (Permadri Pond Coat) will not cure when applied over silicone. Similarly, silicone will not adhere to cured *Zavlar* (Permadri Pond Coat). This incompatibility with silicone is a disadvantage of these products, but can be easily overcome. One strategy is to use butyl rubber, polyurethane caulk, or *3M 5200* instead of silicone when double water- proofing a viewing windows, grommets, and bulkheads.

As an additional measure of security, it is advisable to use a small amount of epoxy to first waterproof the area wherea viewing window or bulkhead will be installed, such that epoxy layer bridges the seam between the *Zavlar* (Permadri Pond Coat) and silicone caulk. This way, in the event the *Zavlar* (Permadri Pond Coat) separates from the silicone caulk,the epoxy provides an additional waterproof layer that will prevent leakage.

Liquid Rubber Summary
PROS: Highly flexible coating resists fractures even in poorly supported tanks. Easy to apply. Minimal smell.
CONS: Limited choice of colors (brown in the case of *Zavlar*/Pond Coat, bright blue in the case of *Ames*). Incompatibility with silicone slightly complicates fish tank viewing window installation and backup waterproofing of grommets and bulkheads.

CHAPTER 11

Pumps & Choosing the Right Pump

Pumps: Overview

Choosing the right pump is often a challenging decision for the newcomers to aquaculture who desire to build their own system. The pump's main purpose is to lift water to a certain height and move a specified amount of water over a unit of time (e.g., gallons per minute, etc.). The perfect mechanical device for this purpose is energy efficient, inexpensive, reliable, and will last a long time with minimal maintenance. **As a general rule, the pump should circulate all of the water from the fish tank every two hours, but preferably closer to every hour, 24 hours a day, seven days a week.**

Make sure the pump you choose will meet this standard of cycling half your system's water every hour. For example, if you have 300 gallons of water in your fish tank, you will need a pump that can move at least 150 gallons per hour (2.5 gallons per minute). An efficient and reliable pump for your aquaculture system is essential. Without a good quality pump that is specifically right for your system, you simply will not succeed.

The flow rate of your aquaculture pump will either be measured in gallons per hour (gph), gallons per minute (gpm), liters per hour (lph) or liters per minute (lpm). **The flow rate will change depending on the head or the height the water must be pumped.** As the head increases the flow rate of your pump will decrease. Be sure to keep this in mind when deciding on your pump. The more head in your system, the more electricity will be required to move a specific amount of water. It takes one pound of pressure per square inch (PSI) to move water up 2.2 feet. The more head, the more pressure required to lift the water.

All pumps are different. One pump may take 100 watts to move five gallons per minute up 10 feet while another might only use 30 watts. The difference is how much energy is consumed and how much heat is produced. Daily power usage equals wattage of pump multiplied by 24 hours. Watt hours = wattage of pump x hours of operation.

Just as you need a powerful enough pump to do the job, make sure you do not get a grossly oversized pump. An oversized pump may move too much water out of your system or overly draw down the fish tank. It will also cause you unnecessary operating expense and/or negative impact the filtration system, causing it to be ineffective. Do not hastily jump in and find the first second-hand submersible pump that you see on the internet. You need the right pump for the specific operational characteristics of your system. Taking the time and doing your due diligence in selecting the right pump in the beginning will save you a lot of headaches later. Your system must operate the way it is supposed to so you will not lose plants and/or fish.

When designing your system, strive to make it as efficient as possible in regard to pumping requirements. Keep the following in consideration:

- The higher the flow rate (amount of water being pumped), the more electricity will be required.
- The higher the water must be pumped (the higher the head), the higher the electricity consumption.
- A timer system will use less electricity than a siphon system as the pump will operate less.
- Keep in mind that the longer the pipe or hose run, the more turns, elbows, bends and fittings that are is the pump outflow line, the greater the friction head (pumping resistance), and thus, the harder the pump will need to work to overcome this resistance.

Steps for Selecting the Right Pump

1. Know the volume of your fish tank.
2. Determine how high (elevation difference) you need to pump the water. This is referred to a 'head', technically speaking. A pump simply capable of moving 40 gallons an hour could be suitable if the tank and the growing beds were adjacent on the same level, but an increase in height and distance between the tank and the growing beds willindicate you need a stronger pump. To calculate what size you need in regard to 'head', measure from the level of the pump intake up to the highest the water needs to be lifted which will be either over the rim of the growing beds, or over the rim of the filter-tank, depending on the system you will be using.

Evaluate the pump's efficiency, and don't be intimidated by this process. It is easy to learn about pump curves.

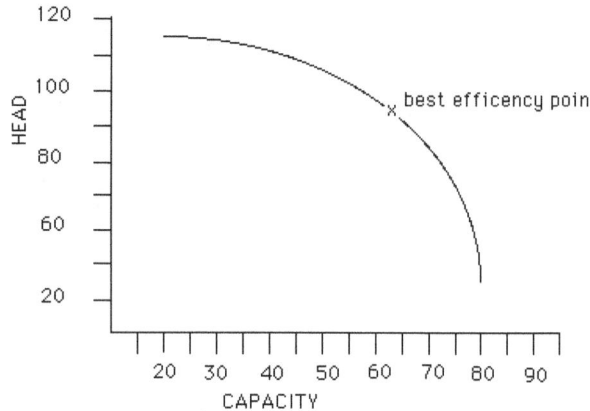

FIGURE 64. *Pump efficiency curve.*

How to Read a Pump Curve:

Figure 64 above shows the head of the pump with its capacity and optimal operating efficiency point. The head of a pump is read in feet or meters. The capacity units will be either gallons per minute, liters per minute, or cubic meters per hour. You can think of the above illustration in metric or imperial units; however you see fit.

The maximum head of this pump is 115 units. This is called the maximum shutoff head of the pump. Also note that the best efficiency point (BEP) of this impeller is between 80 percent and 85 percent of the shutoff head. This 80 percent to 85 percent is typical of centrifugal pumps. The pump manufacturer will typically provide you with the pump curve that will show the exact best efficiency point for the pump in consideration.

Ideally a pump would run at its best efficiency point all of the time, but we seldom hit ideal conditions. As you move away from the BEP, the shaft will deflect, and the pump will experience some vibration. You will have to check with your pump manufacturer to see how far you can safely deviate from the BEP and still get satisfactory operation (a maximum of 10 percent either side is typical).

The pump manufacturer typically states the maximum height the pump can efficiently move the water. Be wary of buying a model that does not supply that information. You may also want to consider using a pump that has a larger head than is strictly necessary for your set-up. This will provide you with the option of diverting the water, should you ever have the need to do so, for maintenance or repair reasons.

Aeration of the water is extremely important and will affect the growth of the fish. Water circulation through pumping action is one way aeration is accomplished (sometimes the only way). However, some operators do use aerators if the pumping action is not providing enough oxygen to the fish. A good efficient pump may cost a little more to purchase, but fish production can be maximized, and operating cost reduced; which can amount to a significant economic advantage over time.

Don't be fooled by advertisers boasting about horsepower, wattage, or voltage. Pump selection should never be made based upon these characteristics. Instead, selection of the pump needs to based entirely on flow rate, head, and pump efficiency. When these parameters closely match your system's parameters, then horsepower, voltage, and wattage will follow.

Customer reviews are helpful. Pumps that are extremely loud or produce a great deal of vibration can impede fish health via stress. User reviews are one of the best ways to determine if this may or may not be a problem with the pump you are considering. **Always have a spare pump on-hand.**

Check your pump regularly to ensure that the inlet is not clogged or partially clogged. As the pump pulls in water, debris can sometimes cling to the inlet hole(s), inlet grate, or pump impeller. Some operators build a screen box for their pump to be housed in to help prevent clogging. Even so, it is wise to inspect and clean the screen box and pump regularly. Maintain some type of back-up power system just in case your electricity is disrupted for an extended time. Imagine having to keep your pump working if the power went off for three weeks. Have an alternative plan in place and the necessary equipment readily available just in case this happens. Provided later in this book will be detailed information on back-up power, associated equipment and alternative energy options.

Shop around for the best pump for the price. There are many places online where pumps can be purchased. Also remember that sometimes we get what we pay for, so be wary of exceptionally low-priced pumps, as well as those on sale at a big discount store or have had the price reduced remarkably. Do your research.

Word of Caution

When buying a pump, or any other product, it is helpful to read product reviews by previous customers. Beware, though, as there are companies with large staffs, mostly located overseas, in which their business is influencing public opinion. For a fee, they will bombard various online retail outlets with positive comments, provide "Likes" or "Dislikes", give a thumbs up or down, and perform other such measures to sway opinion.

Some of the retail, media, and social websites they assail with this service is Facebook, Yelp, mainstream news articles, Yahoo, Google, Amazon, and various other online sites. Many manufacturers, suppliers, and even the federal government, use these companies to push their product or agenda. Coincidently, just before publication of this book I read a solicitation posting on Craigslist offering payment to individuals for posting positive reviews on certain Amazon product listing pages.

Therefore, read product and news article reviews with a skeptical eye. Sometimes it can be more helpful to read the negative product reviews. Obviously, those reviews noted as being from a 'verified purchase' are the most reliable.

PART III : COMPONENTS USED IN AQUACULTURE

CHAPTER 12

Filtration (Mechanical, Biofiltration, Natural)

Filtration

All cultured organisms, vertebrates or invertebrates, finfish, or shellfish, produce waste as a result of the nutrition they receive. Fish excrete ammonia (NH3), mostly from their gills, and it dissolves in the water in which the fish must live. This waste product is toxic to the fish and is an environmental stressor that causes reduced appetite, reduced growth rate, and death at high concentrations.

Maintaining water quality is imperative for fish health. Even with the perfectly designed aquaculture system, it is not a bad idea to have a spare or supplemental filtering mechanism available. If water quality is becoming a concern (regularly or approaches threshold limits on occasion), supplemental filtration can be used (regularly, intermittently to maintain water quality, or just switched on as needed in those rare occasions until water quality reaches optimal conditions). Some aquaculture operators make their own biofilter while others purchase a filtration system.

Natural Filtration

One filtration method that can be incorporated into an aquaculture system is using grow beds: thus, making it an aquaponic system. Please refer to my other books on aquaponics on how to build and operate an aquaponic system. Sized correctly, kept in a favorable temperature range year-round, and plants maintained properly, the flood-and-drain grow beds serve as an ideal natural filtration system. Adding grow beds as a supplemental filtration mechanism can help clean the fish tank while also producing healthy organic vegetables.

Grow beds are addressed in much more detail in my books on aquaponics. However, pertaining to our discussion here, regarding the sole purpose of using grow beds for natural filtration, it is prudent to consider the following guidelines:

- **Size the grow bed using a 1:1 ratio of grow bed volume to fish tank volume.**

The grow bed media should be at least 12-inches deep to allow for growing the widest variety of plants and to pro- vide complete filtration. A grow bed with a freeboard of 1 to 2-inches above the top of the media is helpful in terms of practical application. This allows the operator to work within the media bed without spilling media or water on the floor. Therefore, a media bed 14-inches deep with 12-inches of media is recommended for optimal success.

Design the grow bed to be a working height that best suits the operator. An ergonomically correct height not only reduces strain on the back, arms, and shoulders, but makes the aquaponics experience that much more enjoyable.

Turning over or stirring the bulk of the media within the grow bed every 4 to 12 months is beneficial. If using a liner, be sure not to get too close with your shovel or do anything else that may risk puncturing or tearing the liner.

Use a media guard around all plumbing fixtures. A media guard can be made from a wide variety of common hard-ware supplies, such as: window screen material, larger pipes, wood, aluminum, plastic, paint strainer, etc. A media guard will greatly facilitate cleaning and repair of plumbing fittings. Media guards are also addressed in detail in my aquaponic books.

Mechanical Filtration

The easiest, but most costly, approach is to purchase a filtration unit. The size of filtration unit(s) is dependent upon the volume and operational characteristics of your system (e.g., numbers/size of fish, size of fish tank, etc.).

With so many variables, there is no easy formula for determining the size and number of filtration units needed, or how often they should be used. It is a trial-and-error process. One method, however, is to purchase a filtration unit that is specified to address the volume of water contained within the fish tank.

For instance, assume you have one 400-gallon tank of fish within the normal recommended stocking density range (stocking density is addressed elsewhere in this book), then the following type of filtration unit would be a wise choice. It is designed to for systems up to 400 gallons.

The *Fluval FX6 High Performance Canister Filter* on the following page) has particularly good reviews.

At the time of this writing, it is being sold by Amazon, Petco, Big Als Pets, Pet Mountain, Marine Depot, and other retailers for around $300 USA dollars (2021 prices). The author is not providing a recommendation to any of these merchants, just listing them as options. Furthermore, there are other similar product options which will also work equally as well. The referenced *Fluval FX6* is provided here only as one viable option. Tank volumes in excess of 400 gallons would need multiple *Fulval FX6* units or a larger filter.

Biofiltration

Biological filtration is the use of beneficial bacteria to eliminate organic waste compounds from a body of water. It differs from mechanical filtration in that it is a process whereby water is strained, and suspended material is physically removed from the water.

Adequate filtration (used to trap and remove solid waste) is especially important for aquaculture operations incorporating Nutrient Film Technique (NFT) and Deep Water Culture (DWC) aquaponic systems. **The minimum volume of this biofilter container should be one-sixth that of the fish tank.** Without proper filtration, solid and suspended waste will build up in the channels and canals and will clog the root surfaces. Solid waste accumulation causes blockages in pumps and plumbing components. Finally, unfiltered wastes will also create hazardous anaerobic spots in the system. These anaerobic spots can harbor bacteria that produce hydrogen sulphide, a very toxic and lethal gas for fish, produced from fermentation of solid wastes, which can often be detected as a rotten egg smell.

Biofiltration is the conversion of ammonia and nitrite into nitrate by living bacteria. A lot of fish waste is not filterable using a mechanical filter because the waste is dissolved directly in the water, and the size of these particles is too small to be mechanically removed. Therefore, in order to process this microscopic waste an aquaponic system uses microscopic bacteria.

Fluval FX6 High Performance Canister Filter

Product Overview:
- Super-capacity canister filter for aquariums up to 400 gallons
- Enhanced filter performance for superior aquarium water quality
- Includes all essential filtering media to streamline aquarium filtration

High performance canister filter boasts enhanced performance and setup ease. Fluval FX6 High Performance Canister Filter makes aquarium fishkeeping easier and more convenient. A 1.5-gallon media capacity filter is powered by an energy-efficient pump with a 925-gph pump output that draws just 43 watts. The result is cleaner and healthier aquarium water while reducing operating costs. Fluval FX6 Canister Filter comes with all the essential filtering media to provide complete 3-stage filtration.

Multistage filter features removable media baskets precision engineered to eliminate water by-pass for efficient filtration. Each media basket is lined with a mechanical foam insert for effective mechanical pre-filtering. Instant-release T-handles let you lift and separate the baskets quickly and easily, making routine maintenance simpler. For freshwater or marine aquariums up to 400 gallons.

DIMENSIONS	PUMP GPH	MECHANICAL AREA	FILTER GPH	HEAD HEIGHT	WATTS
17" dia x 21" high	925 gph	325.5 in^2	563 gph	10.8 ft	43W

In an aquaculture system, the biofilter is a deliberately installed component to house a majority of the living bacteria.

The dynamic movement of water within a biofilter will break down exceptionally fine solids, which prevents. Biofiltration is unnecessary in aquaponic media bed systems (flood-and drain, ebb-and-flow) because the grow beds themselves serve as perfect biofilters.

To summarize, fish constantly and continually excrete highly toxic nitrogenous liquid waste from their bodies in the form of ammonia. If ammonia becomes too prevalent in the water, the fish will die.

The ``nitrogen cycle'' (more precisely, the nitrification cycle) is the biological process that converts ammonia into other, relatively harmless nitrogen compounds. Several species of bacteria are involved in this process. Some species convert ammonia ($NH3$) to nitrite ($N02-$), while others convert nitrite to nitrate ($NO3-$). Thus, cycling the tank refers to the process of establishing bacterial colonies in the filter that convert ammonia -> nitrite -> nitrate.

Nitrates are one of the three forms of nitrogen found that plants use to grow and produce chlorophyll and proteins. It is a component of DNA, which transfers genetic information cell reproduction and plant reproduction.

By eliminating the organic waste compounds in the water, biological filtration detoxifies the water and makes it safe for fish. Additionally, by removing the organic waste compounds, algae are controlled because those compounds are the nutriment that algae require in order to grow.

A biological filter or "biofilter" is simply a home for the beneficial bacteria that perform the nitrogen cycle. The filter provides surface area that the bacteria can live on, and a recirculating water pump ensures that water flows over the bacteria so they can obtain the nutrients and oxygen needed for survival.

The bacteria occur naturally in a fish tank or pond. They live on fish and other underwater surfaces. So, if the bacteria occur naturally, then why is a bio filter necessary? It is because fish tanks have a much higher concentration of fish than would occur in nature, and the fish are usually fed a high-protein food. Hence, there is a higher concentration of organic waste (e.g., ammonia) than then naturally occurring bacteria can deal with. A biofilter houses a higher concentration of beneficial bacteria, so that ammonia and the other nitrogen compounds can be reduced to levels which are safe for the fish.

Why do tanks/ponds turn green?

Ammonia is at the root of most green water (algae) problems. As discussed above, ammonia forms naturally in fish tanks and ponds, and is toxic to fish. Ammonia can be reduced or eliminated by the beneficial bacteria in a properly operating bio-filter, but if it is not, the water could become toxic were it not for algae. Algae help protect fish up to a certain point. As ammonia levels rise, algae will colonize the pond or tank by taking up the ammonia as nutrient. However, too many algae makes for an unsafe environment for fish. Some algae produce toxic chemicals that pose a threat to fish. The toxins are released into the water when the algae die and decay. However, algae's biggest threat to fish is its ability to deplete dissolved oxygen in the water; thus, causing the fish to suffocate.

Types of Biological filters

There are two basic types of bio-filters: in-tank/pond and out-of-tank/pond. Out-of-tank/pond filters are divided into two types — pressurized and non-pressurized. The main function of any type of bio-filter (i.e., providing a home to bacteria), is the same regardless of design. The differences

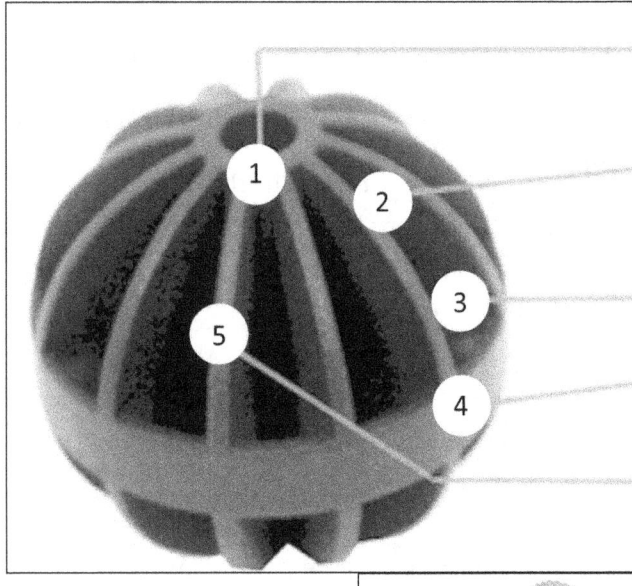

1. Center channel allows the balls to be strung together making cleaning easier.
2. Textured surface maximizes space for beneficial bacteria populations and water retention.
3. Maximizes dissolved oxygen levels.
4. Compact size allows for placement into smaller areas.
5. Paddle wheel design breaks water flow helping to de-gas and aerate water.

FIGURE 65.
Plastic biofilter media material.

are in cleaning, space requirements, and add-on enhancements. Out-of-tank/pond filters work best for aquaponics.

A bio-filter can be a manufactured purchased unit or constructed as a Do-It-Yourself (DIY) project. Both types will be discussed later in this chapter.

The primary purpose of a bio-filter is to provide surface area for beneficial bacteria to live on, the size of the filter depends upon the amount of organic waste the bacteria have to deal with. In aquaponics, the amount of waste is a function of the fish population and uneaten feed.

If the bacteria colony is not thriving, ammonia will accumulate, and algae will colonize the tank or pond. The most common problems with bio-filters are allowing it to be depleted in wastewater, improper cleaning, chlorine or chloramine in the water supply and copper leaching into the water from copper pipes. Winter weather (cold temperatures) can also reduce or kill off bio-filter bacteria.

How to Make a Low-Budget Media Ball Biofilter

The filter is made up of a specially designed inert material such as ceramic or plastic. Plastic is less expensive. Commonly used biofilter mediums are Bioballs®, Kaldnes® Bio Media®, and BioMax Balls®, which are proprietary products available from aquaculture supply stores, aquarium stores, and online. There are many other equally reliable brands on the market, as well. These products are designed to serve as ideal biofilter material, because they are small, specially shaped plastic items that have an exceptionally large surface area for their volume. Other media can be used, including volcanic gravel, plastic bottle caps, nylon shower poufs, netting, polyvinyl chloride (PVC) shavings and nylon scrub pads; however, the plastic bio media balls work best. Any biofilter needs to have a high ratio of surface area to volume, be inert and be easy to rinse. Biofilter balls have almost double the surface area to volume ratio of volcanic gravel, and both have a higher ratio than plastic bottle caps. When using suboptimal biofilter material, it is important to fill the biofilter as much as possible, but even so the surface provided by the media may not sufficient to ensure adequate

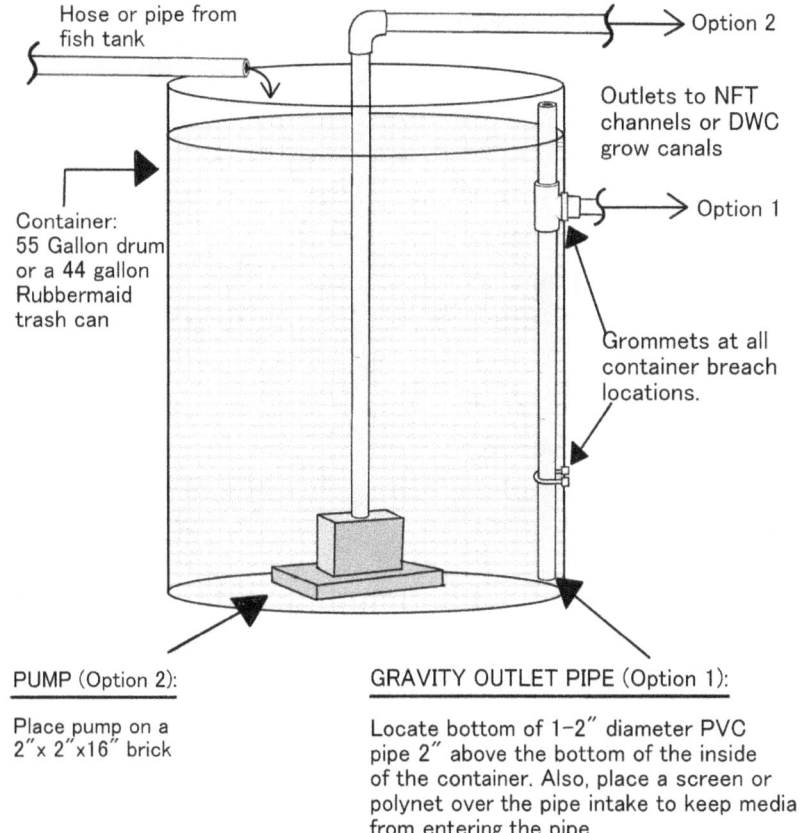

FIGURE 66. Illustration of a Homemade Media Ball Biofilter

biofiltration. It is always better to oversize the biofilter, but secondary biofilters can be added later if necessary.

Biofilters occasionally need stirring or agitating to prevent clogging, and occasionally need rinsed if the solid waste has clogged them, creating anoxic zones. For cleaning, not all the media has to be removed. Removing and power washing approximately 75 percent of the media is usually sufficient. Leaving some media will help jump start good bacteria growth after the cleaned media is returned. Avoid using standard type cleaning solvents. Cleaning solvent residue remaining on the balls can kill the good bacteria and introduce undesired contaminants into the system.

Another required component for the biofilter is aeration. Nitrifying bacteria need adequate access to oxygen in order to oxidize the ammonia. One easy solution is to use an air pump, placing the air stones at the bottom of the container. This ensures that the bacteria have constantly high and stable DO concentrations. Air pumps also help break down any solid or suspended waste not captured by the mechanical separator by agitating and constantly moving the floating bio filter media balls. To further trap solids within the biofilter, it is also possible to insert a small cylindrical plastic bucket full of nylon netting (such as Perlon®), sponges or a net bag full of volcanic gravel at the inlet of the biofilter. The waste is trapped by this secondary filter. The trapped waste is also subject to mineralization and bacterial degradation.

FIGURE 67. *Above View of a Homemade Media Ball Biofilter*

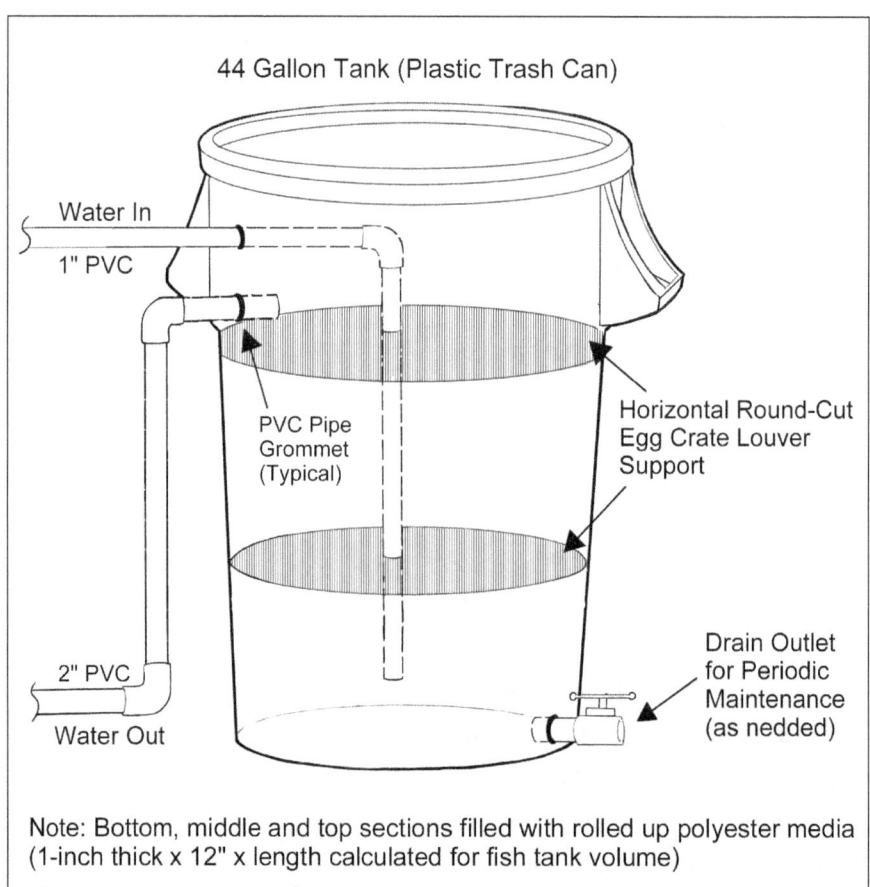

FIGURE 68. *Homemade biofilter*

How to Make a Low-Budget Polyester Woven Biofilter

A biological filter needs to be constructed out of non-corroding material such as plastic, fiberglass, ceramic or rock that has large amounts of surface area nitrifying bacteria cells can colonize. To make bio-filters more compact, material that has a large surface area per unit volume is usually chosen. This unit of measure is usually referred to as the specific surface area (SSA) of the biofilter media. Simply stated, the more surface area available, the more bacteria cells can be grown and the greater the nitrification capacity, which means that higher feed rates can be achieved. A biofilter with a higher SSA will be more compact than one with a lower SSA. Keep in mind, however, that some biofilter media with a higher SSA can become clogged with bacteria. Thus, there must be a balance between a high SSA and an operationally reliable biofilter media.

Below is an illustration of a homemade biofilter, followed by step-by-step instructions.

Steps to making a biofilter:

For systems where additional filtration is needed, then the biofilter can be sized accordingly:

- The ideal rate of filtration is 1/2 gallon of water per minute per square foot of biofilter bed. If your filtration is too fast it will result in insufficient exposure time to the bacteria.
- To calculate the size of your filter, figure the tank volume in gallons and divide by 60 (minutes). This equals the square footage of your biofilter material. This 'area' is the TOTAL area of filter material put it in biofilter unit. Itwill be cut to size and rolled up.
- Acquire the supplies needed:
  ◊ One 44-gallon Rubbermaid (or equal) round trash can. See figure 69 below.

FIGURE 69. Standard 44-gallon plastic drum.

  ◊ A roll of 1-inch-thick HVAC cleanable reusable filter (typically polyester). See figure 70 below.

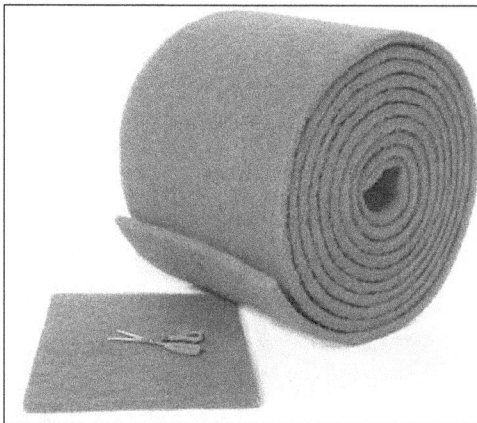

FIGURE 70. HVAC cleanable reusable filter material.

The advantage of using rolled up HVAC cleanable reusable filter material is that in addition to the micro-surface area it provides for bacteria adhesion and growth, it is easy to clean — just unroll it and blast it with water. Use a pressure sprayer to blast it clean if it becomes too clogged for a garden hose to do the job.

Another option is 1-inch-thick dual sided poly fiber filter made of 100 percent polyester specifically for Koi ponds (see figure 72). It is washable or inexpensive enough that it can simply be replaced. It cost approximately $10 for a 12-inch by 120-inch roll (year 2021 USA dollars). It is available online through eBay, Amazon, and various aquarium supply websites. Many local aquarium shops also carry it or can order it for you.

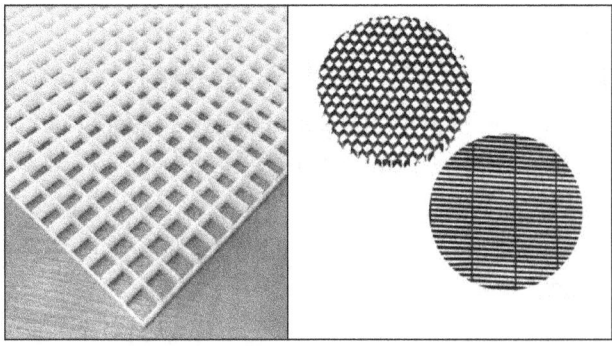

FIGURE 72. Dual sided poly fiber filter made of 100 percent polyester.

FIGURE 71. Dual sided poly fiber filter filter material.

- PVC plumbing pipe and fittings:
  ◊ 3/4" PVC Pipe (2)
  ◊ 2" PVC Pipe (1)
  ◊ 1" PVC Pipe (1)
  ◊ 2" 90 Degree Elbows (2)
  ◊ 2" Male Adapter, for 1/2 of bulkhead adapter
  ◊ 2" Female Adapter, for 1/2 of bulkhead adapter
  ◊ 1" Male Adapter, (2), for 1/2 of bulkhead adapters
  ◊ 1" Female Adapter, (2), for 1/2 of bulkhead adapters
  ◊ 1" 90 Degree Elbows (1)
  ◊ 1" PVC Pipe to Garden Hose Fitting (2)
  ◊ 3/4" 90 Degree Elbows (8), for supports
  ◊ 3/4" Kris Cross Fitting (2), for supports
  ◊ 3/4" PVC fittings to get from the pump to a garden hose on the outside of the box.
- Egg Crate Louver from overhead fluorescent light fixture, for supports. Cut to fit horizontally in the 44-gallon trash can. See photos for various examples.

- Install the plumbing as shown in the DIY illustration below.
- Cut the HVAC cleanable reusable filter area previous calculated.
- Install ¼ of the previously determined HVAC cleanable reusable filter area in the bottom third of the 44-gallon can (this filter material should be rolled up like a newspaper).
- Place a round-cut egg crate louver support horizontally on top of the rolled-up reusable filter material.
- Install ½ of the previously determined reusable filter area in the middle third of the 44-gallon can (this filter material should also be rolled up) on top of the round-cut egg crate louver horizontal support.
- Place the second round-cut egg crate louver horizontally on top of the rolled-up reusable filter material that was placed in the middle section of the can.
- On top of the second horizontally placed round-cut egg crate louver support, install the remaining ¼ of the reusable filter area in the top third of the 44-gallon can (this filter material is also to be rolled up).

Purchasing a Biofilter

Or those aquaculture operators would rather purchase a biofilter than make one, there are many options available. Below is one option that works well for aquaculture systems. The cost for the below biofilter is approximately $80 (USA, year 2021).

TetraPond Clear Choice Biofilter PF1
by *Tetra Pond* (figure 73)

* Works with fish tanks up to 500 gallons.
* Mechanical pre-filter sponges remove suspended debris to improve water clarity.
* Interchangeable .75" and 1" diameter intake fittings.
* Bio Ring media provide large surface areas for beneficial aerobic bacteria.
* Easy out-of-tank accessibility and simple maintenance.
* *ClearChoice* biofiltration removes these contaminants and ammonia using its advanced Trickle Flow and and BioRing technologies.

The *TetraPond ClearChoice BioFilter* works to keep your pond clean through biological and mechanical filtration. This easy-to-use system filters coarse debris and provides biological filtration media with massive surface area for beneficial bacteria to colonize and remove harmful pollutants. The *ClearChoice BioFilter* utilizes a unique Venturi system, which draws air into the in-flowing water, creating ideal conditions for the beneficial bacteria to decompose waste. Take a look at the information below to see how the *ClearChoice BioFilter* could benefit your aquaponic system.

How the TetraPond Clear Choice Biofilter PF1 Works

STAGE 1: Fish water is pumped into the filter. It passes through the foam, which sieves coarse suspended debris from the water (Mechanical Filtration).

STAGE 2: The water from stage 1 then passes through the finer foam, which provides further filtration.

STAGE 3: The final purification process of the water occurs on the surface of the biological filtration media. Beneficial bacteria will naturally colonize on this massive surface area and convert the harmful pollutants in the water into relatively harmless nitrates, which are removed in partial water changes.

The Tetra Venturi and Trickle-Flow System

The water inlet into your filter is fitted with an advanced *Venturi* system, which draws air into the in-flowing water, creating ideal conditions for the beneficial bacteria to decompose waste. This, when combined with the use of "Trickle Filtration" encourages the prolific growth of beneficial bacteria, which helps to keep the pond free of pollutants.

Recommended Pump

TetraPond recommends the *TetraPond Water Garden Pump* 325 GPH or 550 GPH. Do not use pumps with flows that exceed 550 gallons per hour unless a flow control valve is used.

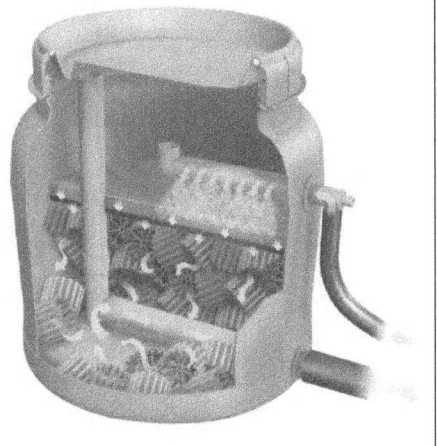

FIGURE 73. TetraPond Clear Choice Biofilter PF-1.

PART III : COMPONENTS USED IN AQUACULTURE

CHAPTER 13

Alternative Energy Options & Operating Off-The-Grid

Alternative Energy Overview

Renewable energy is clean, affordable, domestic, and effectively infinite. It produces no emissions and results in cleaner air and water for all. Energy prices are rising rapidly, and fuel oil prices are unpredictable. With the increasing cost of food, gas, utilities, and taxes consuming more of our budget, consumers are spending less on non-essentials. It is no surprise that many families and business owners are wondering how they will survive. Two strategies that help greatly consist of lowering energy consumption by becoming more energy efficient and by using alternative energy sources.

In recent years, a lot of attention has been given to biomass as an alternative fuel source (ethanol, aka corn). With many Americans having to sustainably curtail their food budget and hundreds of millions around the world starving, U.S. lawmakers have adopted an insane policy of burning up our food supply in the form of a corn-based ethanol fuel mandate. What is really crazy is that ethanol-laced fuel gets much worse mileage than gasoline; you have to buy more of it to get where you are going. It is a policy that has never made much sense, but adopted because of lobbyists, Big Agriculture, and political favors. Obviously, growing corn for ethanol reduces the available farmland to grow food crops. Furthermore, a recent Congressional Budget Office report concluded that the increased use of ethanol accounts for 10-15 percent of the increase in food prices; thus, one of the reasons food prices continue to rise, despite more efficient farming practices being implemented and more land being converted to agriculture.

Solar and wind energy are alternative energy options that not only make good sense but are relatively affordable and easy for most people and small businesses to put into place. This chapter will address those options, and provide other helpful information to help you lower operation costs and enable you to more energy independent.

Economics of Electricity

The cost of electricity can be somewhat challenging to determine, as most utility companies have different rates for summer and winter months, and for peak and not-peak usage. With the development of so called 'smart meters', utility companies are able to determine exactly when and how much electricity you are using.

Before we see how much electricity costs, we have to understand how it's measured. When you buy gas, they charge you by the gallon. When you

TABLE 19. **Measuring Electricity Costs**

DEVICE	WATTAGE	HOURS USED	kWh
Medium Window-Unit AC	1000 watts	one hour	1 kWh
Large Window-Unit AC	1500 watts	one hour	1.5 kWh
Small Window-Unit AC	500 watts	one hour	0.5 kWh
42" Ceiling Fan on Low Speed	24 watts	ten hours	0.24 kWh
Light Bulb	100 watts	730 hours (all month)	73 kWh
CFL Light Bulb	25 watts	730 hours	18 kWh

buy electricity, they charge you by the *kilowatt-hour (kWh)*. When you use 1,000 watts for 1 hour, that's a kilowatt-hour.

To get kilowatt-hours, take the wattage of the device, multiply by the number of hours you use it, and divide by 1,000. (Dividing by 1,000 changes it from watt-hours to kilowatt-hours.) Here's the formula to figure the cost of running a device:

wattage x hours used ÷ 1000 x price per kWh = cost of electricity

For example, let's say you leave a 100-watt bulb running continuously (730 hours a month), and you're paying 15¢/kWh. Your cost to run the bulb all month is 100 x 730 ÷ 1,000 x 15¢ = $10.95.

If your device doesn't list wattage, but it does list amps, then just multiply the amps times the voltage to get the watts. For example:

2.5 amps x 120 volts = 300 watts

(If you're outside North America, your country probably uses 220 to 240 volts instead of 120.)

Watts vs. Watt-Hours

- Watts is the *rate* of use at *this instant*. In other words, Watt is a measure of work.
- Watt-hours is the total energy used over time.
- We use *watts* to see how hungry a device is for power. (i.e., 100-watt bulb is twice as hungry as a 50-watt bulb.)
- We use *watt-hours* to see how much electricity we used over a period of time. That's what we're paying for.

So, just multiply the *watts* times the *hours used* to get the *watt-hours*. (Then divide by 1,000 to get the kilowatt-hours, which is how your utility charges you.) Example: 100-watt bulb x 2 hours ÷ 1,000 = 0.2 kWh.

The national average rate for electricity is useless for two reasons:

1. **Electricity rates vary *widely*.** They vary not only by region (i.e., an average of 7.5¢ in Idaho vs. 36¢ in Hawaii), but they also vary from the same utility. As a matter of fact, rates can range from 12¢ to 50¢ per kWh from the same provider. The only way to know what you're actually paying is to check your bill carefully.
2. **Electric rates are usually *tiered*,** meaning that excessive use is billed at a higher rate. This is important because your *savings* are also figured for the highest tier you're in. For example, let's say you pay 10¢/kWh for the first 500 kWh, and then 15¢/kWh for use above that. If you normally use 900 kWh a month, then every kWh you save reduces your bill by 15¢ (technically, once you get your use below 500 kWh, then your savings will be 10¢ kWh, but you get the point).
3. **Electric rates typically vary throughout a 24 hour period.** Higher rates are assessed during peak usage hours and lower rates during times when electricity usage is lower.

For simplicity in determining a 'rough' cost for an item, use a rate of 15¢ per kWh. This isn't a "typical" rate, since there's no such thing as typical when it comes to electricity rates. And it's certainly not average. It's just a *reasonable expectation.* Your own rate could be dramatically higher or lower than 15¢ per kWh

Furthermore, fees and taxes are added to the base electrical charges. It is easy to see how even a small solar powered system can help alleviate some of this financial burden, whether it be used for aquaponics or for other purposes.

Electrical Definitions
- Watt'—a measure of work.
- Kilowatt (kW or kw)—a unit of power, equal to 1,000 watts.
- Kilowatt-hour (*kWh* or kW·h)—a measure of electrical energy equivalent to a power consumption of 1,000 wattsfor 1 hour.
- Ampere (amp)—a measure of electricity in motion.
- Volt—a measure of electricity under pressure.

Grid Connection

Depending on local regulations and laws, so-called net-metering systems can be installed that allow your electricity meter to run 'backwards' when your power generator is producing more than you need. In some locations, excess powerdelivered to the grid will result in additional reimbursements. In many states, electrical utilities are mandated to purchase a specific percentage of their energy from 'green' or renewable sources and welcome special arrangements with energy producers.

Solar Energy Fundamentals

Solar power is one of the fastest growing sectors in the U.S. The price of solar panels has dropped by 60 percent since 2010 and costs continue to fall. This is not just a trend in America, the rest of the world is also moving in the same direction. While plants convert sunlight into biomass production, photovoltaic (PV) panels convert it into electricity. The conversion efficiencies of PV panels have increased over the years to as high as 17 percent at maximum light intensity. Some experimental PV cells have achieved efficiencies of 40 percent.

PV panels should be mounted for maximum light interception. In the Northern hemisphere, panels can be attached to south-facing roofs, other support structures, or on a tracking device that follows the position of the sun across the sky. PV panels generate DC power that can be converted to AC power to operate lights, pumps, and other household/ small business equipment. PV systems can be interconnected with the local electrical grid, ensuring that electrical power is always available. In a grid-connected system, excess power from the solar installation can be sent to the grid.

Interconnection requirements vary from state-to-state and utility-to-utility. Off-grid PV systems require some form of electrical storage to provide power during periods of little or no sun. Typically, banks of batteries are installed for this purpose. Off-grid systems are best suited to applications where there is no nearby electrical grid or for standalone systems, such as aquaculture.

A significant portion of sunlight reaches the surface of the Earth as heat radiation. This energy can be used to heat water. Typically, not much water is needed for washing and cleaning purposes, but aquaponic operators can use warm water to heat the greenhouse. The most solar energy that can be collected is during the middle of the day, so storing the warm water for use during the night is a good strategy to reduce the use of heating fuel. A large fish tank can serve as an efficient means of regulating temperature in a greenhouse or other enclosed area. The rise in heating prices makes long-term storage of warm water more attractive.

There are also many other technologies for converting incoming solar radiation into heat. The most common systems are flat plate collectors that allow water or other fluids to flow through a panel that is oriented toward the sun. Very simple flat plate collectors are often used for heating swimming pools but can and are used for many other applications. Flat plate collectors are very efficient. Other systems for converting sunlight into heat include evacuated tube collectors and parabolic reflectors. These products are capable of generating higher temperatures but are significantly more expensive and are often dependent upon a tracking system to maintain an optimal orientation.

Other solar technologies are passive. For example, big windows placed on the sunny side of a building allow sunlight to heat-absorbent materials on the floor and walls. These surfaces then release the heat at night to keep the building warm.

Benefits of Solar Energy Summarized

- Feel good about saving energy and the environment.
- Helps promote a model for sustainable living.
- Save tens of thousands of dollars over the life of your solar energy system.
- Allows you to put current utility bill money to better use.
- Use all your appliances without feeling guilty.
- Helps the U.S. get one step closer to energy independence.
- Doing something good for future generations.
- Solar energy is an inexhaustible fuel source that is pollution and noise free.

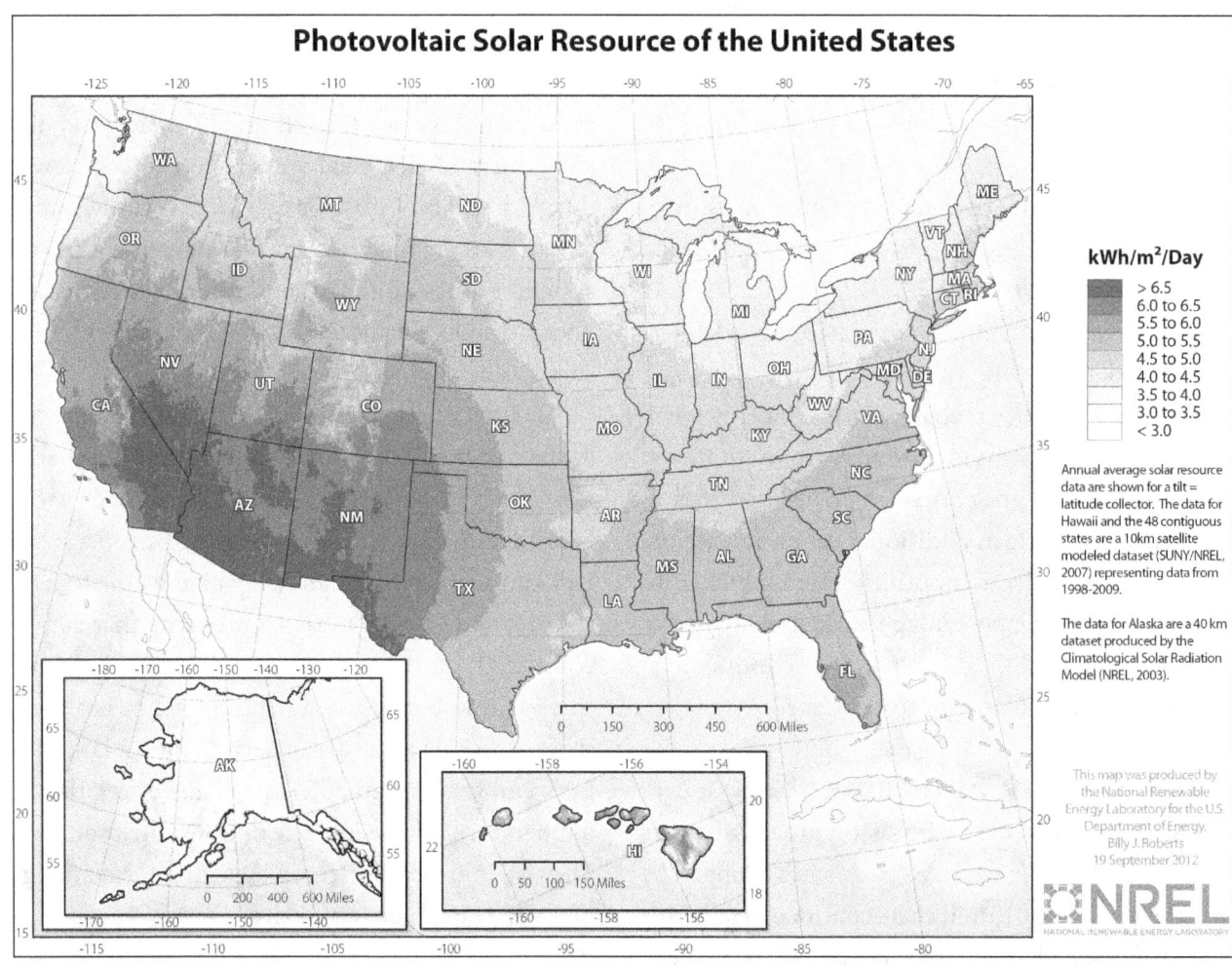

FIGURE 74. *The American Solar Energy Society provides additional information on their web site: (http://www.ases.org/)*

Average Daily Solar Radiation per Month at Your Location

Figure 74 shows the general trends in the amount of solar radiation received in the United States and its territories.

Steps to Integrating Solar Energy

1. Review your power needs.
2. Look for ways that you can conserve power (e.g. more efficient lighting, temperature control, etc.).
3. Calculate the savings over time, and consider how those funds could be better spent.
4. Calculate the cost of installation.
5. Remember that it is an investment, and that there is a good probability that you can sell the system (thus receiving a portion of your money back) in the future if you ever decide to go a different direction.
6. Obtain free advice from solar energy equipment vendors regarding your operation.
7. Educate yourself on solar energy equipment.
8. Compare cost.
9. Installation.
10. Maintenance (clean the panels periodically).

Solar Energy Grants, Loans, Tax Incentives, and Other Resources

- Energy Technology Inc. has solar powered equipment, supplies, and accessories for homeowners and small businesses: http://personal-solar.com
- Solar Resource Calculator at:
- http://solar.ucsd.edu/SolarApp.html
- The Solar and Energy Loan Fund (SELF) provides low-cost financing options and energy expertise to help homeowners and small business owners lower energy bills and implement alternative energy systems, such as solar equipment:
- http://cleanenergyloanprogram.org/how-it-works/homeowners
- The USDA Rural Business and Cooperative Programs has a wide array of programs, as well as grants and loans, for increased energy efficiency and alternative energy systems: http://www.rurdev.usda.gov/Energy.html
- Check with your electric utility company to see if they have other resources as well as provide solar energy conversion benefits.
- With tax incentives, solar electricity typically pays for itself in five to ten years.

Wind Energy

The United States has some of the best wind resources in the world, with enough potential energy to produce nearly 10 times the country's existing power needs. Wind energy has become a cost-effective source of new-generation power technology for residential dwellings. Upfront capital cost has dropped steadily over the past few years, as wind turbine technology has improved. Currently, over 400 American manufacturing plants build wind components, towers, and blades. Using wind to pump water and generate power is not a new idea. Before the start of rural electrification in 1936, wind energy was widely used across the U.S. During the last ten years, technological improvements and rising energy prices have significantly increased the number of wind energy installations. In many cases, large installations have occurred on farmland, but the farmers are not the main users of the generated energy, nor do they own the equipment. Many farmers only receive a land lease payment for the area used by wind turbines.

The success of wind energy installations depends on site-specific wind conditions. Wind maps have been compiled for all regions of the U.S. and these maps are useful for a first approximation of the average wind speed at a given location (links provided at the end of this section). However, local topology, vegetation, and building structures significantly affect the average wind speed. Where possible, use local wind speed measurements to determine whether a site is appropriate for wind generation. Currently, an average wind speed of 9 mph for small wind generators and 13 mph for large generators (measured at 100 ft above the ground) is considered necessary for the economical use of wind power.

Even small wind generators can be used to operate a fairly large size aquaponic system. Such a system can be operated off-grid, but a connection to the grid is prudent. Grid-connected-systems have the advantage of having power available when the wind system is not functioning at full capacity and does not require batteries for electrical storage.

Wind Energy Helpful Resources
- Remarkably interesting current wind map of USA:
- http://hint.fm/wind
- A collection of helpful worldwide wind and climate maps:
- http://www.climate-charts.com/World-Climate-Maps.html
- The U.S. Department of Energy provides an 80-meter (262.5-ft) height, high-resolution wind resource map for the United States with links to state wind maps. States, utilities, and wind energy developers use utility-scale wind resource maps to locate and quantify the wind resource, identifying potentially windy sites within a fairly large region and determining a potential site's economic and technical viability. http://www.windpoweringamerica.gov/wind_maps.asp

Grant, Loan and Rebate Programs

Local utilities, as well as state and federal organizations, offer a variety of grants, loans, and rebate programs for alternative energy installations. Each of these programs comes with its own set of requirements and often entail cost-sharing. Nevertheless, these programs can reduce the investment costs and/or reduce the pay-back period. Many of these programs are announced on web sites, requiring some effort to learn about them. In some states, energy regulating commissions, such as the Board of Public Utilities or state energy agencies, have programs for renewable energy systems. Your local utility and county extension service, state departments of agriculture, the USDA, and the NRCS are good places to start investigating the various opportunities.

Renewable Energy Certificates

Some states administer renewable energy certificate (REC) programs that allow certified producers of eligible renewable energy to sell these certificates that represent proof that 1,000 kWh of electricity was produced. Thus, in addition to reducing your electric power consumption from the utility grid (e.g., by lowering your monthly electricity bill or receiving payment for excess electricity you exported to the grid), the RECs generated by your system can provide additional income when sold (i.e., to a power company that was mandated to deliver a certain percentage of its total output as renewable energy). While prices for RECs fluctuate, REC programs provide additional financial incentives for renewable energy production (http://www.eere.energy.gov/greenpower/markets/).

Energy Conservation

Before you consider a solar or wind system for your aquaculture operation, the first step in any renewable energy project is ensuring that the existing system is functioning efficiently. The reason is quite simple: **the cost of implementing energy efficiency measures is less than the cost of installing renewable energy technologies to compensate for inefficient use of conventional energy sources.**

Energy Conservation Resources
- An excellent energy conservation guide full of valuable ideas and practical techniques has been produced by the U.S. Dept. of Energy and can be found at http://energy.gov/sites/prod/files/2013/06/f2/energy_savers.pdf.

- The State of California Consumer Energy Conservation Department has a list of energy conservation and efficiency tips for your home, office, business, vehicle, and other areas, located at: http://www.consumerenergycenter.org/tips/index.html
- A useful reference source is the book titled "Energy Conservation for Commercial Greenhouses", published by NRAES.
- Some of the many ways to achieve better energy performance in greenhouses include using thermal curtains, where possible, and checking that they seal properly (i.e., form a continuous barrier), verifying that environmental control systems are doing what they are supposed to, and sealing glazing leaks through unintended openings in your walls and roofs.
- Most local utility companies have a brochure and/or webpage showing you many ways on how you can better con- serve energy in your home and business.

Living or Operating Off-The-Grid

A growing trend, especially within the United States, is being "off-the-grid". This term can conjure up many different meanings for people, ranging from a peaceful community of folks who still travel by horse-and-buggy to people just wanting to live exclusively from sustainable farming practices to groups preparing for the end of the world. While these groups can fall under the basic definition of off-the-grid, the general term is simply people who are not connected to a public utility. Their reasons for choosing to live in such a way varies from one end of the spectrum to the other. For the most part, off-the-grid folks are not connected to their local electrical system; they are a stand-alone-system.

Off-The-Grid Alternative Energy Source

Electricity, however, is just the main aspect of being off-the-grid; it can also relate to all energy sources. For instance, off-the-grid homes are autonomous; they don't depend upon the municipal water supply, sewer, natural gas, cable or internet services, or similarly related utility services. A true off-the-grid house is one that is able to operate completely independent of all traditional public utility services.

Although there are many different types of alternative energy sources (such as geothermal), the ones that are most applicable to the topic-at-hand are wind, solar, biofuel, ethanol, and water (well, creek, or pond). Of these, solar polar is by far the most commonly used alternative energy source.

The initial set-up costs for living off-the-grid are a bit high, which discourages many from considering alternative energy. However, a large percentage of people find that the initial price tag is compensated through time as they are free from the monthly financial drain as well as government mandated taxes and regulations. In addition, many have peace of mind knowing that they are generating a non-polluting energy source and that they won't have the deal with as many outages and shortages. Lastly, they take comfort in knowing that they are no longer financially supporting and sanctioning the fossil and nuclear fuel industries.

Alternative Food Source

Many who live off-the-grid have taken to growing their own food as well. Compared to taking the jump into an alternative energy source, growing your own food will require constant dedication and a lot of hard work; the end results are always worth it, though. Growing your own food has many advantages, such as:

- **Receiving the nutrients your body requires.** Much of the food we eat today has been modified and changed with chemicals and preservatives.

PART III : COMPONENTS USED IN AQUACULTURE

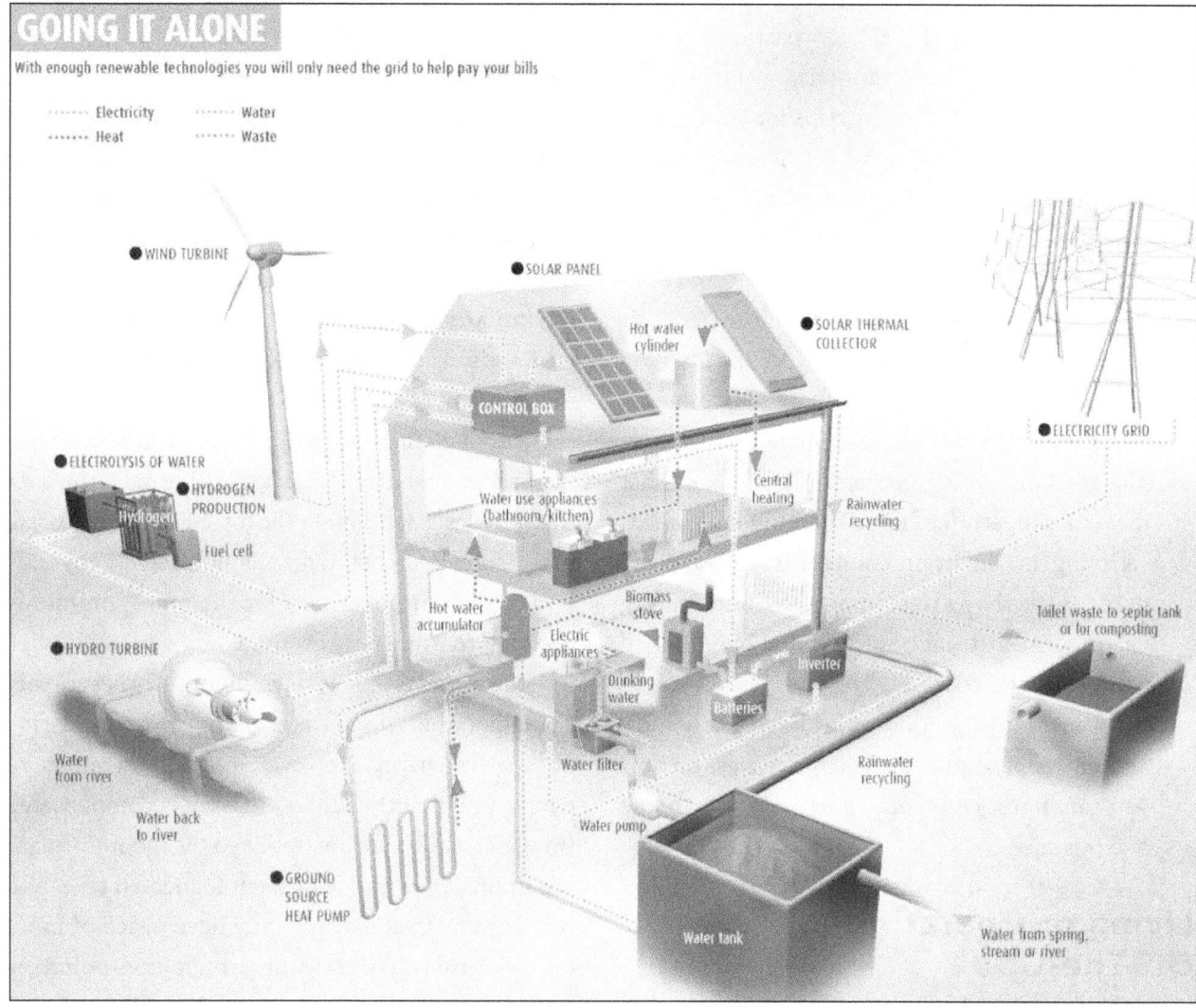

It only takes about ten minutes of research online will quickly learn that of what we is being sold at the grocery store just isn't healthy to consume. Most of the chemicals used in farming were approved by the Environmental Protection Agency (EPA) without any research into how these chemicals could harm individuals. Currently, the EPA considers 60 percent of all herbicides, 90 percent of all fungicides, and 30 percent of all pesticides as carcinogenic (cancer-causing). A study conducted by the National Academy of Sciences found that pesticides may contribute to an additional four million cancer cases in the United States alone. Imagine what that number might be today! Our body craves particular minerals, nutrients, and vitamins; growing and eating your ownorganic crops helps meet this need. Bonus: the food tastes better than anything you will buy from the store.

- **It will save you money!** The math is simple: if you are no longer spending money every week at the grocery store, butare growing your own food to consume, then you are saving money that will rapidly add up over time. If you decide to sell some of it to neighbors, family, and friends, then you just created an additional cash flow as well.
- **Stops soil erosion.** According to The Soil Conservation Service, over three billion tons of topsoil are eroded each year from the U.S.'s

croplands; or about seven times faster than it is being built-up by mother nature.

- **Better water quality.** Due to most crops across the United States being sprayed with an array of chemicals (such as pesticides), at least 38 states have reported that their groundwater has been contaminated. A possible outcome of drinking such water is cancer. Growing your own crops, free of pesticides and other chemical agents, helps prevent the polluting of your own groundwater (which you may be drinking if you are truly off-the-grid).
- **Saves energy.** Most crops you buy at the store today are grown on mega-farms. In order to keep up with production, many farmers are forced to use petroleum (more than any other single U.S. industry). This energy is used to create the synthetic fertilizers as opposed to growing healthy crops.
- **Emergency Supply.** Should a disaster ever strike your family or community, whether that be from a storm or some-thing worse, you will have a ready-supply of food to consume. Many know what a grocery store looks like right before a major storm system hits; shelves are emptied in a hurry leaving you with few options if you didn't arrive first (and if you did, you may have left with multiple bruises and a higher blood-pressure). The old Boy Scout motto of "always be prepared" is a good one to remember here.
- **Makes good sense:** While it takes a lot of work, time, and effort, the benefits of being off-the-grid and supplying your own power and food far out-weight the negative. Upfront costs for setting up are eventually reimbursed, and there is a peace-of- mind that simply cannot be purchased when you are no longer dependent upon local utility companies, stores, and certain government regulations.
- **Friendly Reminder:** As mentioned previously in this chapter, one should be sure that making alternative energy changes and/or implementing a farm (e.g., adding solar panels, harnessing the wind, growing a garden, etc.) doesn't violate any local laws, will not void your homeowners insurance, will be tolerated by your local utility companies, and is conformance with your homeowner association (if applicable).

PART III : COMPONENTS USED IN AQUACULTURE

PART IV

Operation and Maintenance

PART IV : OPERATION AND MAINTENANCE

CHAPTER 14

Starting, Operating, and Troubleshooting Your Aquaculture System

Successfully Activating Your Aquaponic System

When starting the system, you must be very patient. Don't expect your system to be running smoothly immediately. It can take time for your fish to get acclimated and in a thriving state of development. This is a good time to gain experience in testing the water with a test kits and getting familiar with your automatic monitoring systems (should you elect to install such). Practice testing the tap water immediately out of the faucet (if you are using a public water source) and test it again after it has been in the system every couple of days. Also practice recording the water quality readings, adjustments to your system that you have made, and organizing everything associated with your system (e.g., supplies, paperwork, tools, etc.). This will make things go much more smoothly over the long run, saving you time and trouble in the future. Most municipal tap water has chlorine added. This water needs to be dechlorinated. When starting, the water can be dechlorinated in the system over a few days. After your system it is running, it is important to dechlorinate the water before adding fish to your system by letting it sit in another container for a day or two (if you have another container available). The more water, the longer it will take it to dechlorinate. If your tap water contains chloramines, it will be necessary to obtain a chemical neutralizer, available in most fish shops or online, to properly address this issue (just follow the directions on the container).

After you put your system together and fill it with water and run the system for a few weeks as if it were fully stocked. This will provide you with an opportunity to troubleshoot any problems and should be enough time to let chlorine dissipate from the system. It will also allow you time to evaluate whether or not you need to rearrange your system for more optimal growth conditions. Ensure all equipment (i.e., timers, backup power system, pumps, etc.) and piping is functioning properly.

During the startup process it is important to monitor the water quality closely, as well as the state of health of the fish. At this initial stage of getting your system started, toxic ammonia and nitrites are accumulating in the tanks. This can be incredibly stressful to the fish, and even be toxic if not maintained properly.

When starting the system, it is best to feed the fish small amounts of food to avoid ammonia spike and algal bloom. Cease feeding the fish if you observe an algae bloom or the ammonia/nitrites levels increasing to a high level of 1 ppm; as this signals that the bacterial population is insufficient to transform the wastes. Only resume feeding when the ammonia and nitrite levels return to 0.5 ppm and below. Start increasing the feeding rates after two months, when a thriving bacterial base has been established.

System Startup Checklist

1. Define your short-term and long-term goals. What is your objective for getting into aquaculture? Where would you like to be with your aquaculture endeavor in five years?
2. Decide on the size of system to build.
3. Draw design plans. Take into consideration a plan for future potential expansion.
4. Research where to get parts, equipment, and fish.
5. Buy and assemble components.
6. Fill system with water and circulate (at least a week).
7. Add fish to the system. Start out with about 20 percent of the final maximum capacity stocking density.
8. Monitor water quality and perform partial water changes as needed.

How to Introduce New Fish into Your Tank

Moving into a new home can be a shock for any creature, but for fish it is particularly stressful. Not only will a new tankbe completely unfamiliar for your new fish, but the water will likely be of a different quality. Subtle changes in water temperature, pH, and other factors make a big difference in a fish's life. To keep your fish safe and healthy, there are certain measures you should take when introducing fish to a new tank. Following are some tips on how to make an easy transition:

1. Test your water quality before adding your new fish. Your chlorine level should be at zero, and your pH should match that of where the fish is coming from.
2. Gradually add your system water to the transport tank to equalize the temperature.
3. Feed the fish in your system before adding any new fish. This will make the existing fish less aggressive.
4. Before adding your new fish, dim the lights. This will create a less stressful environment.
5. When the transport tank water has relatively the same water parameters (i.e. temperature, pH, dissolved oxygen)as your fish tank, then relocate the fish to the new fish tank. Only relocate the fish. Do not relocate the transport
6. If possible, add more than one fish at a time, as this reduces the chance of one fish being picked on and harassed.

Tips and Tricks for Aquaculture

1. Maintain system. Stay organized. Keep good records.
2. Always test the pH of the system. This will let you know if your system will have a tendency to be more on the acid orbase side, so that you can be better prepared in regard to taking appropriate measures in maintaining the ideal pH.
3. For smaller systems vitamin C and an air pump can be used to bubble out chlorine and chloramine more quickly from tap water.
4. Worms, such as red wigglers, in media beds (if integrating a aquaponics into your system) can be used to breakdown solids and reduce anaerobic zones.
5. Never use cleaning products, pesticides, algaecides, fertilizers or like substances in fish tanks.

6. Avoid direct sunlight on fish tanks. You can limit light on the fish tank to avoid algae growth. Although Tilapia will eat algae.
7. Never change more than 1/3 of water at a time. More than that will destroy the good bacteria in the system.
8. Make sure you have backup power available for pumps and aerators. Purchasing an inverter box (approx. $25 USD from an automotive store) and connecting it to an automobile battery when the power is out is the least expensive way to have a backup power. supply.
9. Prepare for system failure. Be proactive by having at least one spare pump and battery backup system on hand.
10. It is prudent to keep spare dechlorinated water readily on hand. It is recommended that you store at least 10 percent of your system water volume in a separate container. If using municipal water be sure to let it sit for a minimum of 24 hours before adding to your system so it can properly dechlorinate. Water will need to be added to the system on a regular basis to offset evaporation. Adding chlorinated water directly to your system will kill off the beneficial bacteria.

Aquaculture System Operation and Maintenance

Following is a preview of the tasks that can be expected in aquaponics It may seem like a lot of information to keep up with, but really, once the system has been established and you learn what to do, it only takes a small amount of your time. Furthermore, there are automated monitors and controls that you can integrate into your system which will simplify the process even more.

Visual Inspection

A daily visual inspection is necessary to ensure that the system is operating as it should. A walk-through inspection would consist of observing the filters, water clarity, pumps, and aerators, for any problems. The water circulation between the filter and fish tanks should be flowing normally. Water levels should be adequate, and topped off, if necessary, especially during the summer.

Water Quality and Temperature Monitoring

An entire chapter in this book is dedicated to water quality (Chapter 15). The Water Quality chapter provides more detail than what is covered in this section. However, this section will provide a brief overview on water quality and temperature. Maintaining water quality is essential to the success your aquaculture operation. Water pH should be checked at least 3–4 times a week. The pH levels should be maintained in the range of 7 to 9 range for Tilapia. This pH range is optimal for Tilapia.

If the pH falls below the optimal range, you can alternate between using calcium hydroxide and potassium carbon- ate or potassium hydroxide. Both calcium and potassium are important minerals for healthy plant growth. To bring down the pH levels, you can use some of the same pH regulators that are used in hydroponics. Be sure to avoid using pH regulators that contain sodium, as they are detrimental to the life within your system. In addition, do not use products containing citrus to adjust pH.

The water introduced into the system should be free of chlorine and chemicals. Bacteria colonies will be killed off by chlorine, chemicals can be harmful to all life within the system and enter into your food chain. If using municipal water, it is important to let it stand in a separate tank for 2 to 3 days so the chlorine can vaporize off before being added to your system. Aerating the water in the tank can speed up the process.

Also monitor the water temperature using a submersible thermometer to ensure that water in the fish tank is kept stable and within the optimal range of your fish. Depending on your location, and the particular species of fish you are raising, you may need additional heating and insulation to help maintain temperatures.

Feeding of Fish

Feeding your fish is often viewed as one of the most enjoyable daily tasks by aquaculture operators. Fish need to be fed 1-2 times a day, but they don't necessarily have to be fed at regular timings. The process can even be automated with the use of auto-feeders. Some of these feeders can be acquired fairly inexpensively.

Most fish used in aquaculture can be fed with commercial feeds, worms that you even grow yourself, or with another species, scraps of food. Some operators have separate tanks that they use to grow algae or duckweed in which they net out periodically to feed their Tilapia. Beware of commercial fish feed that is not organic if you are desiring to maintain a food supply that is completely free of heavy metals and other chemicals used on crops, as GMO feed will be a gateway of such into your food chain.

When fish are not eating, it typically signifies that they are stressed. It could be that they are too cold, too hot, there is too much direct lighting, oxygen is lacking, and/or there are water quality problems. If this happens, it is important to check all system parameters to identify the problem and correct it right away.

Routine Aquaculture System Management Practices

Below are daily, weekly, and monthly activities to perform to ensure that the aquaculture system is running well. These lists should be made into checklists and recorded. That way, multiple operators always know exactly what to do, and checklists prevent carelessness that can occur with routine activities. These lists are not meant to be exhaustive, but merely a guideline for management activities based upon a typical aquaponic system. Following this section are two 'go-by' checklist forms which will hopefully be helpful.

Daily Activities

Check that the water, filter, and air pumps are working well and clean their inlets from obstructions.
* Check that water is flowing.
* Check the water level, and add additional water to compensate for evaporation, as necessary.
* Check for leaks.
* Check water temperature.
* Feed the fish (2–3 times a day if possible), remove uneaten feed and adjust feeding rates.
* At each feeding, check the behavior and appearance of the fish.
* Remove any dead fish.
* Remove solids from the fish tank and rinse filters.

Weekly Activities

* Perform water quality tests for pH, ammonia, nitrite, and nitrate before feeding the fish.
* Adjust the pH, as necessary.
* Check the fish tank from the bottom up.
* If using a biofilter, check it.
* Check that nothing is obstructing pipes, filters, pump(s) and water flow.

Monthly Activities

* Stock new fish in the tanks, if required.
* Clean out the biofilter, or other filters, if not operating efficiently.
* Clean the bottom of the fish tank using fish nets.
* Weigh a sample of fish and check thoroughly for any disease.

TABLE 20. **Aquaculture System Data Tracking Sheet**

DATE	pH	AMMONIA	NITRATE	NITRITE	NOTES

System Maintenance Overview

1. Feed the fish daily, monitor fish health.
2. Test water quality (every other day for the first month, then about once a week, and when you suspect problems. As needed clean out filter screens, filter tanks (if using), tubing, water pump, etc.

Aquaculture Work Tools

Following are two immensely helpful tools for the system operator:

1. Aquaculture System Data Tracking Worksheet (table 20)
2. Daily Maintenance Checklist (table 21)

TABLE 21. **Aquaculture System Maintenance Checklist**

Month_____

	MONDAY	TUESDAY	WEDNESDAY	THURSDAY	FRIDAY	SATURDAY	SUNDAY
Week 1							
Week 2							
Week 3							
Week 4							
Week 5							

Weekly Tasks

	WEEK 1	WEEK 2	WEEK 3	WEEK 4
Check pH				
Check Ammonia				
Add Water				
Check for Insects				

Monthly Tasks

	MONTH
Clean Out Pumps and Plumbing	
Check Nitrate Levels	

Harvesting of Fish and Stocking of New Fish

With the right set-up, and when done correctly, aquaculture allows for the regular harvesting of fish. You do not have to harvest the entire tank of fish at the same time. In fact, it is not advisable to do so. As a matter of fact, the closer you keep the system operating at optimal stocking density, the better. You can stock the tank with new fingerlings when the fish population has gone below the optimal stocking density. Since Tilapia will, under certain conditions, eat their young it is important that you follow proper protocol, described later in this book, when adding fry or fingerlings to the tank or pond.

Fish Health Check and Addressing Fish Illness

It is prudent to regularly net some fish to check on their state of health. Also, consider having a separate tank to isolate fish that appear to be sick, so as to prevent the spread diseases among the fish population.

Some aquaculture operators maintain a separate quarantine fish tank for sick fish. They will pull sick fish from their system and put them in the quarantine tank. They then add salt to the quarantine tank periodically for disease control. Salt acts as a natural anti-bacterial agent on the fish body. Due to various risk to the overall fish health, this practice needs to be done slowly over time, with water quality and fish health closely monitored throughout and following the process. Only pure sea salt or swimming pool salt should be used. Table salt should never be used.

It is important to add only predetermined calculated amounts of salt as different fish species have varying levels of tolerance. Too much salt can stunt plant growth, or even be fatal to the plants and fish. Whenever salt is added, it is important that a refractometer be used throughout the process so salt concentration can be properly monitored. Do not add salt to the aquaculture system.

TABLE 22. **Potential Emergencies**

(not all situations are applicable to all different types of aquaponic systems)

TYPE OF PROBLEM	CAUSES
Beyond your control	Flood, tornadoes, wind, snow, ice, storms, electrical outages, vandalism/theft
Staff error, diffusers plugged	Operator "errors", overlooked maintenance causing failure of back-up systems or systems components, alarms deactivated.
Tank water level	Drain valve opened, standpipe fallen or removed, leak in system, overflowing tank.
Water flow	Valve shut or opened too far, pump failure, loss of suction head, intake screen plugged, pipe plugged.
Water quality	Low dissolved oxygen, high CO_2, supersaturated water supply, high or low temperature, high ammonia, nitrite, or nitrate, low alkalinity.
Filters	Channeling/plugged filters, excessive head loss
Aeration system	Blower motor overheating because of excessive back-pressure, drive belt loose or broken or disconnected, leaks in supply lines.

Addressing Emergencies

Emergency Response Processes to Have In Place
- Ready Phone List
  - Plumbers
  - Electricians
- Trained Staff
  - Logical Troubleshooting Procedures
  - Triage
- System Designed to Fail Reliably
  - Staff
  - Fire/Emergency
  - Know when and who to call for addition help.
  - Safety

Necessary Response Times for Emergencies
- **High** (fast response time — minutes)
  - electrical power
  - water level in tank
  - dissolved oxygen — aeration system/oxygen system
- **Medium** (moderate response time — hours)
  - temperature
  - carbon dioxide
- **Low** (normally slowly changing — days)
  - pH
  - alkalinity
  - ammonia-nitrogen
  - nitrite-nitrogen
  - nitrate-nitrogen

Aquaculture Safety

Safety is important for both the operator and the system itself. The most dangerous aspect is the proximity of electricity and water, so proper precautions need to be taken. Food safety is also important. Proper protocols need to be maintained to ensure that no pathogens are transferred to the fish or introduced into the human food chain. Finally, it is important to take precautions against introducing harmful chemicals to the system.

Electrical Safety

Always use a residual-current device (RCD). This is a type of circuit breaker that will cut the power to the system if electricity grounds into the water. The best option is to have an electrician install one at the main electric junction. Alternatively, RCD adaptors are available, and inexpensive, at any hardware or home improvement store. An example of an RCD can be found on most hairdryers. This simple precaution can save lives. Moreover, never hang wires over the fish tanks or filters. Protect cables, sockets, and plugs from the elements, especially rain, splashing water, and humidity. There are outdoor junction boxes available for these purposes. Check often for exposed wires, frayed cables, or faulty equipment and replace accordingly. Utilize "drip loops" where appropriate to prevent water from running down a wire into the junction. Never leave extension cords on the ground where water escaping the system can come into contact with it (such can be fatal to human life).

Food Safety

Good agricultural practices (GAPs) should be adopted to reduce as far as possible any food-borne illnesses, and several apply to aquacultures. The first and most important is simple: always be clean. Most diseases that affect humans would be introduced into the system by the operators. Use proper hand-washing techniques and always sanitize harvesting equipment. When harvesting, do not let the water touch the produce; do not let wet hands or wet gloves touch the produce either. If present, most pathogens are in the system water and not on the produce. Always wash produce after harvesting, and again before consumption.

Do not place harvesting equipment on the ground. Prevent vermin, such as rats, from entering the system, and keep pets and livestock away from the area. Warm-blooded animals often carry diseases that can be transferred to humans. Prevent birds from contaminating the system however possible, through the use of exclusion netting and deterrents. If using rainwater collection, ensure that birds are not roosting on the collection area, or consider treating the water before adding it to the system. Preferably, do not handle the fish with bare hands; disposable gloves are always the safest bet.

General Safety

Often aquaculture systems, farms, and gardens in general, have other general hazards that can be avoided with simple precautions. Avoid leaving power cords, air lines, or pipes in walkways as they can pose a trip hazard. Water and media are heavy, so use proper lifting techniques. Wear protective gloves when working with the fish and avoid the spines. Treat any scrapes and punctures immediately with standard first-aid procedures—washing, disinfecting, and bandaging the wound. Seek medical attention, if necessary. Do not let blood or body fluids enter the system, and do not work with open wounds. When constructing the system, be aware of saws, drills, and other tools.

Keep acids and bases in safe storage areas and use proper safety gear when handling these chemicals. Always keep all dangerous chemicals and objects safely stored and away from children. Take precautions to ensure children cannot fall into the pond or fish tank.

Safety Summary Overview

- Take all necessary safety precautions where electricity and water combinations occur.
- Follow all National Electric Codes.
- Use RCD on electric components to avoid electrocution.
- Use only low voltage equipment and sources 24 VAC or 12 or 24 VDC.
- Shelter any electric connections from rain, splashes, and humidity using correct equipment.
- Implement a maintenance schedule & regular system checks.
- Ensure that everyone working with the system is properly trained.
- Have strict policies, procedures, and supervision involved for all working with the system, especially for entry level staff and visitors.
- Follow OSHA guidelines for all aspects of the operation.
- Do not contaminate the system by using bare hands in the water.
- Avoid trip hazards by keeping a neat workstation.
- Wear gloves when handling fish and avoid spines.
- Wash and disinfect wounds immediately. Do not work with open wounds. Do not let blood enter the system.
- Be careful with power tools and dangerous chemicals; wear protective gear.
- Adopt GAPs to prevent contamination of produce. Always keep harvesting tools clean, wash hands often, and weargloves. Do not let animal feces contaminate the system.

Water Quality Troubleshooting: Potential Problems & Solutions

Problem: *The pH over 9.5 or pH is less than 6.5 (or out of your systems optimal range)*
Reason: Acidity of water is too low or too high.
Solution: The pH ranges for optimal growth for Tilapia are between 7 and 9. Refer to section in this book regarding pHand how to adjust it. Remember though, quick pH level changes are hazardous to fish. Since most systems naturally settle into their own operating pH level over time, adjusting and maintain pH beyond this natural range can be an ongoing challenge. If a system component is already buffering the system, additives will only change pH temporarily. It is best to find long-term solutions.

Problem: *Low Water Level*
Reason: Surface evaporation reduces water in system. Warmer temperatures, low humidity, and turbulent water will result in greater evaporation. Ensure there are no water leaks.
Solution: Top off system with dechlorinated water.

Problem: *Water is Green*
Reason: Too many algae. May be due to water having too many nutrients and/or too much light hitting the water surface.
Solution: Make sure you are not over feeding the fish, and the fish tank is properly shaded. Also note that some algae eating fish will only eat certain types of algae.

Problem: *Water is Dirty*
Reason: Usually a result of fish not eating all of their food, and/or the system is under-filtered.
Solution: Ensure that you are not over feeding. Use a towel or sock as a filter on the return water to help filter out sediment that is not captured in the grow bed.

Problem: *Water is Cloudy*
Reason: Overfeeding, system is out of balance, algae growth, and/or the system is under-filtered.
Solution: Ensure that you are not over feeding. Also make sure that all water parameters (i.e. temperature, pH, dissolved oxygen, etc.) are in the desired range.

Problem: *Top of Tank Has Iced Over*
Reason: Conditions are too cold.
Solution: Adjust temperature of water and/or room temperature. If fish are still alive, add air bubbles to fish tank to maintain oxygen levels.

Problem: *Water is Foaming*
Reason: Detergents or other chemicals may have been introduced into the system.
Solution: Perform 50 percent dechlorinated water changes every day until the foaming is gone. It is important to prevent fish shock as much as possible in this situation, but it is also critical to eliminate the contaminants.

Fish Troubleshooting: Potential Problems & Solutions

Primary Ingredients for Fish Health:

1. Know and maintain the optimal water parameters for your fish species (e.g., temperature, pH, dissolved oxygen, etc.).
2. Feed fish at a rate appropriate to the biomass of your fish and stage of growth. Observe feedings to ensure fish are being fed enough or that there is not a lot of waste. Adjust feedings appropriate with growth rate and with changes of stocked density.
3. Shade fish tank as appropriate. No direct lighting on fish tank.
4. Remove or minimize stress (e.g., vibration, loud noises, banging, etc.). The author prefers a system configuration that does not have the pump inside the fish tank.

Problem: *Dead Fish*
Reason: Unknown.
Solution: Remove dead fish from the tank immediately to prevent decaying ammonia buildup. Try to identify the reason for death. Observe other fish to ensure there is not major problem with the system.

Problem: *Fish Seriously Sick, Almost Dead, No Hope of Recovery*
Reason: Unknown.
Solution: Euthanasia. Don't flush, suffocate (by allowing fish to die out of water), freeze, or use antacid. These methods are forms of extended torture. A more humane approach is to remove the fish from the tank and flatten fish's head hard and fast with a brick.

Problem: *Fish Swimming at Surface of Water Unusually or Gulping Air at Surface ('piping')*
Reason: Lack of dissolved oxygen. However, piping can also be a symptom of fish being infected with gill parasites.
Solution: Immediately add an air bubbler. Manually splash water but try not to stress fish. Water/air exchange introduces oxygen into the water. Test to see if pH levels are low and increase slowly if necessary. Remember, changing pH as little as 0.2 quickly can be dangerous to the fish.

Problem: *Fish Acting Unusual, (e.g., not behaving normally, swimming sideways, floating upside down, etc.)*
Reason: Possibly a problem with the feed you are using. Could be the result of a disease, injury, or poor water quality. Odd behaviors could indicate breeding activity.
Solution: Monitor activity. Try a different feed. Do an internet search with the specific symptoms and type of fish. Consult a fish expert.

Problem: *Fish Jumping Out of Tank*
Reason: It could be that the water level is too high, they are trying to catch insects, a result of poor water quality, or water parameters are not within the proper range for your fish species.
Solution: Reduce water level. If necessary, cover tank with wire mesh/netting, test water parameter conditions, test waterquality, and/or filter water intake with a rag or sock to catch sediment if water is not clear.

Problem: *Fish Has Creamy White Film Slime*
Reason: Could be the result of low pH level, breeding behavior, or disease.
Solution: Check pH level, try to determine if it is potentially disease related (fish not acting healthy), and do an internetsearch with the specific symptoms.

Problem: *Fish has been determined to have Ich, Ick, or White Spot Disease*
Reason: A protozoa (one-cell animal parasite) called Ichthyophthirius multifiliis. Typically takes hold as a result of stress (environment change/fighting) which often negatively impacts the immune system of the fish making it more susceptible to such problems.
Solution: Easily treated if caught in time. It is best to treat the fish in a separate quarantine tank. Other solutions include adding one-tablespoon sea salt (non-iodized salt) per 1-gallon water in the system. In addition, gradually raise water temperature to the upper limits of your fish species optimal temperature range.
Purchase one of the following products online, or from your local aquarium or pet shop: *Coppersafe*, *Quick-Cure Ich-Ease*, *Aquari-sol*, *Cure-Ick*, or *Super Ick Cure*. Follow directions.

Problem: *Fish Disappeared*
Reason: Critter, thief, or the fish flopped out of tank.
Solution: Cover tank with wire netting and secure it, erect a scarecrow to the tank, close and lock down area at night, and/or monitor aggressive fish.

Problem: *Algae on Fish*
Reason: Algae does not grow on fish; it grows on fungus on your fish. It can also be the result of cyanobacteria and/or high nitrates in the tank. Too much light on the fish tank can exacerbate this problem.
Solution: Ensure tank conditions are in the optimal range. Use a secondary filter method filter the intake (e.g., a towel or sock -- something extra to filter the water before if enters back into the fish tank).

Problem: *Fish Are Not Eating in Colder Weather*
Reason: Fish eat less the colder the water temperature. Water temperature impacts fish metabolism. Fish digestive enzymes are also not as active when temperature drops.
Solution: Feed less and/or raise the temperature (depending upon the circumstances and species). Every situation is unique.

Problem: *Power Outage*
Reason: Power grid is unreliable.
Solution: Always have a backup system readily available.

Problem: *Water Pump Stops*
Reasons: Lack of electricity, defective or clogged pump.
Solution: A system crash can occur in less than 24 hours. Switch to backup power system (if power outage). Replace pump right away. It is prudent to have at least one back-up pump readily available.

Problem: *System is Leaking Water*
Reason: Parts are old and cracking, sealants/tape defective.
Solution: Temporarily seal holes in pipes with *Teflon* plumbers tape, replace worn parts, shutoff water flow to parts of system that are leaking and repair, use a silicon product to seal leaks.

Problem: *Red Worms Appeared (Aquaponics Grow Bed)*
Reason: Probably introduced by natural elements, feed, or someone else.
Solution: Do nothing, as red worms provide additional benefit, and can be good free fish food.

PART IV : OPERATION AND MAINTENANCE

CHAPTER 15

Water Quality

Fish Life Support

As already addressed in previous chapters of this book, growing fresh fish from your own aquaculture system has many advantages. To summarize key points of these previous chapters, even small systems can produce such high production yields that excess fish can be given as gifts, bartered, and/or sold for a profit. Most importantly, you can rest assured that the fish that you are eating is free of mercury and other toxins, unlike the fish typically sold in grocery stores and served at restaurants.

Because fish, are totally dependent upon the water surrounding them and are unable to escape this environment, the quality of the water is of great importance. Several water quality parameters are important in fish culture, including dissolved oxygen (DO), temperature, mineral content, and the concentrations of several waste products excreted by the organisms. The mere presence of fish has a direct impact water chemistry. For example, fish excrete waste products that are high in ammonia and other nitrogen compounds, which tend to pollute the water. Consequently, the relationship between the fish in your aquaculture system and the surrounding water is a dynamic and crucially important one.

Tilapia Environmental Requirements

Tilapia can tolerate a wider range of environmental conditions—including factors such as salinity, dissolved oxygen, temperature, pH, and ammonia levels (table 23)—than most cultured freshwater fishes can. In general, most tilapia are highly tolerant of saline waters, although salinity tolerance differs among species. Nile tilapia is thought to be the least adaptable to marked changes (direct transfer, 18 parts per thousand in salinity); Mozambique, blue, and redbelly (T. zilli) are the most salt tolerant. With the exception of Nile tilapia, other tilapia species can grow and reproduce at salinity concentrations of up to 36 parts per thousand, but optimal performance measures (reproduction and growth) are attained at salinities up to 19 parts per thousand.

Tilapias are highly tolerant of low dissolved oxygen concentration, but optimum growth is obtained at concentrations greater than 3 mg/L. Temperature is a major metabolic modifier in these fish. Optimal growing temperatures are typically be-tween 22° C (72° F) and 29° C (84° F); spawning normally occurs at temperatures greater than 22° C (72° F).Most tilapia species are unable to survive at temperatures below 10° C (50° F), and growth is poor below 20° C (68° F).

Blue tilapia are the most cold tolerant, surviving at temperatures as low as 8° C (46° F), while other species can tolerate temperatures as high as 42° C (108° F).

Other water quality characteristics relevant to tilapia culture are pH and ammonia. The best growth rates for tilapia are achieved between pH 7 to 9. Ammonia is toxic to tilapia at concentrations of 2.5mg/L. Optimum ammonia concentrations are estimated to be below 0.05 mg/L.

TABLE 23. **Limits and Optimal Water Quality Requirements for Tilapia**

Parameter	Range	Optimum for growth
Salinity, parts per thousand	Up to 36	Up 19
Dissolved oxygen, mg/L	Down to 0.1	> 3
Temperature, C°	8–42	22–29
pH	3.7–11	7–9
Ammonia, mg/L	Up to 7.1	< 0.05

NOTE: It is important that the beginner not be intimidated or overwhelmed by the thought of maintaining a balanced aquaculture system. If constructed and maintained according to the basic principles and parameters stated in this book, operating a successful aquaculture system is a relatively simple and very enjoyable endeavor.

Fish Population Density

Aquaculture stocking density is usually defined as the weight of fish held in regard to the volume of water. The accepted approach for stating fish stocking density is the kilograms (weight) of fish per liters (volume) of water.

The more fish, the greater yield, and the higher profit. However, stocking too many fish will lead to ill health of the fish. The fish will not grow as well and will start dying, and there will be a buildup of wastes which will be toxic to both fish. Finding the perfect balance of stocking as many fish as possible without adverse impacts is key.

Some species of fish are able thrive in a denser population while others require considerably more space. Some species are territorial and expend a lot of energy fighting if packed too close together. Whereas others prefer to be in a denser pack, shoaling, and schooling. Territorial fish forced to live close together expend a great deal of energy, have high stress levels, and lower survival rates. In addition, it is not ethical to raise animals under stressful conditions. As a rule, smaller fish are more likely to live out their lives in schools, although some large fish will school together. Tilapia do fine when stocked at the recommended densities.

What is the recommended stocking density? It will depend upon your filter capacity, the feeding rate, ammonia production rate, oxygen consumption rate, and aeration rate. Either gas exchange or ammonia toxicity are what will most likely be the limiting factor that will determine your stocking density. An experienced operator is generally able to stock a tank at 1 lb. of fish per 5-10 gallons of water, or one to two fish per 10 gallons of water (38 liters). Inexperienced operators are better off following a stocking density of one fish per 22.5 gallons (85 liters) of water. However, it is more accurate to talk about stocking densities in terms of kilograms of fish per cubic meter or liters of waters, as this is the industry standard and doing math via the metric system is a much easier process. For most aquaculture set-ups, using 66 to 88 pounds of fish to 264 gallons (30 to 40 kg of fish per 1000 liters) will work fine.

Large, commercial recirculating fish growing facilities will keep fish at densities up to approximately 3-lbs/ft^3 (50 kg/m^3) in aerated systems and up to 9.5-lbs/ft^3 (150 kg/m^3) or higher in direct oxygen injected systems. Aquaculture hobbyists rarely approach these high fish densities in their smaller-scale systems; however, this does not mean that the standards of fish tank design should not be adhered to. In order to ensure fish health and well-being via water quality optimization, most hobby aquaponics practitioners keep fish at densities well below 1.5-lbs/ft^3 (20 kg/m^3). Commercial aquaculture operators will often reach higher aerated fish stocking densities as this is required to recover capital and operating costs, and so the design and efficient operational characteristics of the system are critical.

Lower stocking densities are much more manageable and have higher success rates as there is more room for error. Lower stocking densities are recommended for those that are new to aquacultures. The amount of food you put into the tanks will also dictate how well the system runs. If you feed too much, there will be an accumulation of detrimental waste which can overwhelm the filtration system.

Ideally, you will harvest the fish as soon as they are big enough. This avoids ending up with a stocking density which otherwise strains the system. Since individual fish grow at different rates, you should start taking out the bigger ones as soon as they are large enough to eat, rather than waiting until you can harvest all the fish at one time. This approach works best for smaller operations, as you will have an ongoing manageable harvest rate where you can eat and/or sell them without having to freeze storage a large quantity of fish at one time.

Temperature Range

What is the temperature range of your system location? Ideally, it should match the optimal temperature of your fish. If not, you will incur higher upfront capital cost and more expensive ongoing operational cost maintaining the temperature at the preferred range of the fish, via heating, cooling, and insulation adjustments. Also, a temporary power failure can end up being catastrophic to your system if it has to be kept at a significant different temperature than the surrounding environment.

Fish prefer a specific temperature range and do best when raised at their optimal temperature. Placing a fish tank in the garage, a shop, or greenhouse will help maintain a stable water temperature.

Fish cannot regulate their body temperature like humans do. They are dependent on the water temperature for their body temperature. To maintain fish health the water temperature should never be adjusted, or allowed to fluctuate, more than 3°F per day.

Rates of Growth

Obviously, fast growing fish provide food and revenue quicker than slow growing fish. However, fast growing fish have higher metabolic rates, consume more food, and produce more wastes. This is not necessarily a bad thing. It just means that growing conditions need to be monitored more closely, and adjustments made to accommodate these higher growth rates.

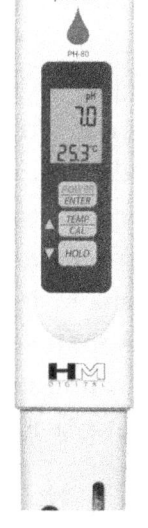

FIGURE 75.

HMDPHM80 Digital pH/Temp. Meter
- Measures pH and Temperature
- One-touch automatic digital calibration
- Simultaneous temperature display
- Water resistant
- Cost: $45 (USA, year 2020)
- Cost and quality of pH/Temperature meters vary considerably from approx. $20 to $300+. This meter, or similar from another manufacturer, is sufficient for most aquaculture and aquaponic operations.

Balance of Water Constituents

One also needs to be careful not to assume that fish data that is based upon a particular species is applicable to all fish species. Each species, and even varieties of certain species, will have unique characteristics, growth rates, and require specific growing conditions.

The fish tank encompasses a direct relationship between ammonia levels, water temperature, and pH, and if their inter-relationships are not understood it can lead to impending disaster.

Dissolved oxygen levels also plays a particularly important role. The following need to be well understood by the operator:

Ammonia

All fish give off ammonia. Ammonia is generated from their gills and waste. Uneaten fish food also is converted into ammonia as it breaks down. If not properly addressed the buildup of ammonia becomes toxic to fish. Ammonia levels as small as 5 ppm is typically deadly to fish.

An overstocked tank with fish is much more susceptible to fatal flaws if any of these parameters are pushed beyond their tolerable limits. Ammonia levels will depend greatly on the temperature of the fish tank water and the pH of the water, fish density, and how much uneaten fish food remains floating in the system. It gets more complicated with warmer water and a pH that is out of balance. Nevertheless, it is easy to monitor these conditions, and there are a number of things that can be done to ensure the system remains healthy and successful. Ideally your ammonia level should be nearzero but there will always be traceable amounts being emitted constantly by your fish. It gets more complicated if you stock your system with a heavy fish load.

Fish can tolerate higher levels of ammonia the cooler the water. The same goes for dissolved oxygen. Cold water canstore more dissolved oxygen than the same volume of warm water.

Understanding the relationship between ammonia and water temperature provides you with the ability to better manage your aquaponic system and avert catastrophic danger.

pH Range

The pH is the measure of the hydrogen ion (H+) concentration in the water. The pH scale ranges from 0-14 with a pHof 7 being neutral. Pure water has a pH of 7, but additives (i.e., chlorine, fluoride, etc.) can alter the pH. A pH below 7 isacidic and a pH of above 7 as basic. Different species of fish have varying pH requirements. However, the ideal pH range for Tilapia is between 7 and 9.

The pH levels of tap water can significantly exceed the required pH range required. As a matter of fact, such detrimental differences are fairly common. Therefore, the pH of tap water needs to be properly adjusted before adding the fish.

Adjusting pH: Adjust pH Slowly After Fish are Present

When fish are already in the tank, it is important to adjust the pH slowly. If pH is lowered too much too fast, it will stress the fish, and can even be fatal. Therefore, pH should be lowered over the span of a week or more when fish are present.

pH Fluctuations

The pH will change in response to system input (rain, fish, plants, topping off the tank periodically). Temperature will also cause your pH to vary. When testing pH, measure at several points in the day for an average, or measure at the same time and temperature each day in order to obtain reliable data.

In the beginning, pH will fluctuate significantly. This is normal. Adding water directly from public water sources is not a good idea because most municipal water has chlorine, which is toxic to the fish and beneficial bacteria. It will cause the pH to change. Capture and store municipal water in a separate

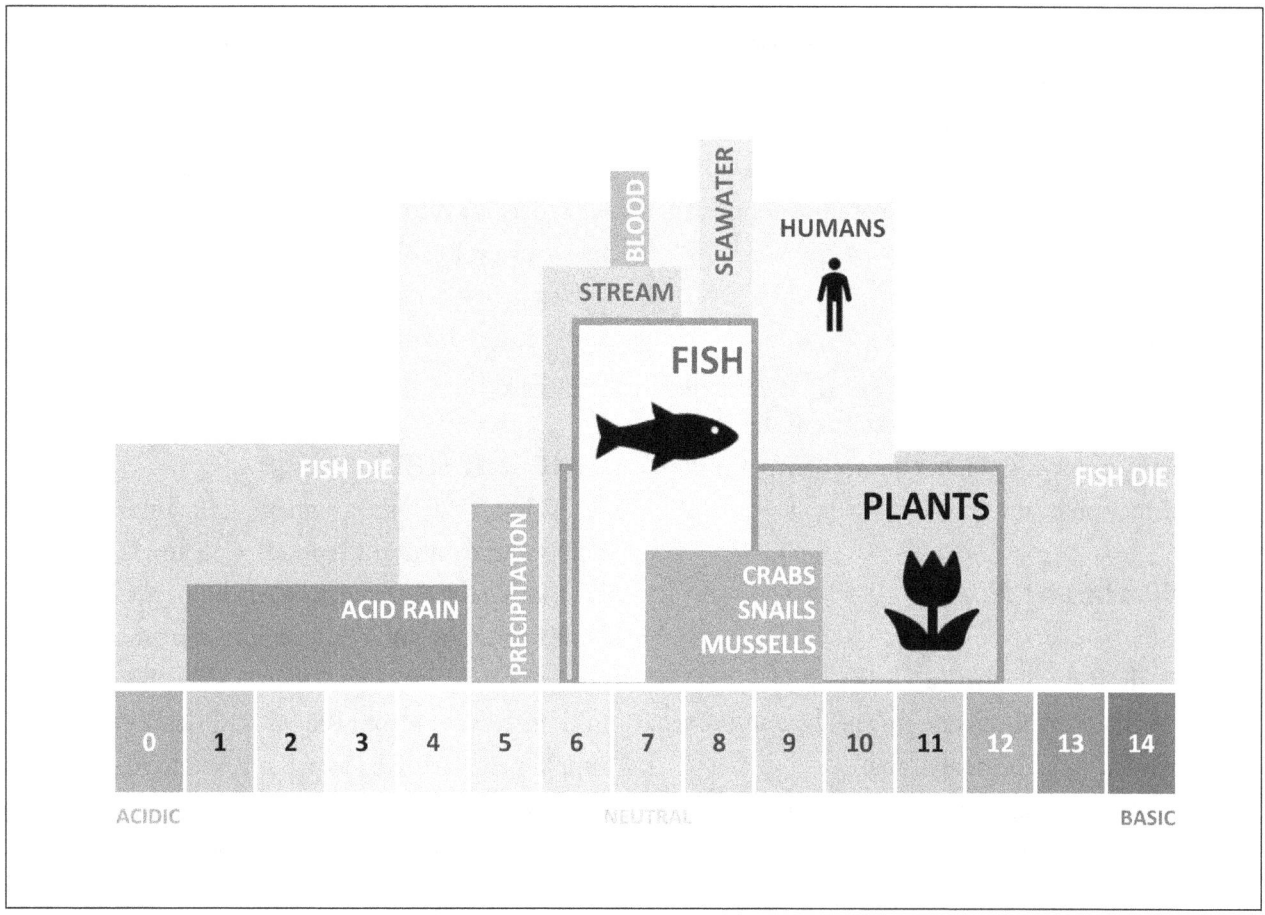

FIGURE 76. *Typical pH range of various living organisms.*

container (e.g., 55-gallon drum, etc.) for a few days, which will allow the chlorine to 'gas off', before adding the water to your system.

Lowering pH

Again, the lower the pH, the more acidic the water. The following are some safe additives that can be implemented to lower the pH in your fish tank.

* Hydrochloric Acid one or two caps per 250 gallons
* Acetic Acid (Vinegar)
* Sulphuric Acid
* Maidenwell media or Diatomite
* Iron sulfate fertilizer

Hydrochloric acid (swimming pool acid) is most frequently used to lower pH to the optimal level. Some gravel, such as limestone, will lower pH. Injecting CO_2 directly into the water has also been reported to lower pH.

Citric acid should never be used. Citric acid is an antibacterial agent that can be fatal to the good bacteria living within your system.

Raising pH

Higher pH readings are called "base". The following are some additives that can safely be used in an aquaculture system to raise the pH:

* Dolomite Lime — Calcium Magnesium Carbonate
* Calcium Hydroxide (hydrated/builder's/slaked/hydrated limes)
* Potassium Carbonate (bicarbonate)
* Potassium Hydroxide (pearl ash/potash)
* Snail Shells
* Seashells
* Egg Shells

If using shells, it is best practice to boil, bleach, or use hydrogen peroxide on them first in order to kill all bacteria. Containing these chemicals or ingredients in a nylon stocking, paint strainer sack, or other breathable bag will allow for easy removal once the desired pH range is achieved.

pH Monitoring

Regular monitoring and recording pH enables one to not only evaluate the current status of the system at a specific point in time, but shows any pH trends in one direction or another.

pH Adjustment Notes:

1. Adjusting pH fast can be hazardous to fish. A matter of fact, changing pH by only 0.2 too quickly can be dangerous.
2. If your system has a pH crash, identify the problem and remedy what caused it rather than just trying to adjust the pH when starting over.
3. Small, crushed particles work faster while larger particles work slower to maintain system pH.
4. Keep "pH increase" equivalents on hand: sodium bicarbonate (baking soda), limestone, and calcium carbonate (eggshells, snail shells, seashells).
5. It is easier to increase pH than it is to decrease.
6. Keep "pH decrease" equivalents on hand: vinegar, Hydrochloric Acid (one or two caps per 250 gallons), Acetic Acid (Vinegar), Sulphuric Acid, Maidenwell media, Diatomite, or Iron sulfate fertilizer.
7. If possible, keep a backup water tank (or barrel) nearby filled with dechlorinated water bucket for topping off the tank periodically as water is lost through evaporation and plant transpiration, or even in an emergency (i.e., pipe leak, etc.). Preferably, this back up supply will hold at least 10 percent of the fish tank volume.

pH Impacts Upon Fish Breeding

There is a relationship between pH and aquatic life breeding cycles. As pH changes, so will the number of fish eggs that are produced, if they can be produced at all. Keep this in mind if you desire to reproduce your fish.

Calcium Carbonate

What is calcium carbonate? Calcium is a mineral that is found naturally in foods. Calcium is necessary for many normal functions of the body, especially bone formation and maintenance. Calcium can also bind to other minerals (such as phosphate) and aid in their removal from the body. Calcium carbonate supplements are used to prevent and to treat calcium deficiencies. Natural forms of calcium carbonate are eggshells, snail shells, and seashells.

Even though calcium carbonate works to lower pH, and natural forms don't do any harm, be cautious in using it. It can take a long time to dissolve and affect the pH over a lengthy time period. Therefore, after adding a little, wait and check pH after two hours before adding any more. Adding too much at one time or over a short period can cause the pH to spike and could result in doing more harm than good.

How much should you add to your system to raise the pH? That question does not have a specific answer. It all depends on the type of calcium carbonate you are using and its form (i.e., natural eggshells, bag of ground powder, etc.), the size of your system, fish density, the type of media you are using (if incorporating an aquaponics grow bed into your aquaculture system), temperature, and how far your pH is off the optimal range.

It is always better to be safe than sorry. Raise the pH using small amounts of calcium carbonate over time, little by little every two hours. With some experience, you will come to know your system, and have a better understanding of how much to add whenever the pH needs to be adjusted.

Dissolved Oxygen (DO)

The limit of dissolved oxygen in fish tank water is termed "saturation level" and the level is measured in parts per million (ppm). DO is essential to the health of your fish and can be impacted by the temperature of your water. Beneficial nitrifying bacteria, which break down the metabolic waste of aquatic creatures, are equally dependent on oxygen.

Oxygen is introduced into the aquaculture system naturally through waves, cascades, and turbulence. If these natural methods are not sufficient, then an aeration device is necessary. Poor aeration in your tank, and/or an extremely high density of fish in your tank can lead to inadequate DO.

The amount of oxygen that water can hold is temperature dependent. Warm water holds less oxygen than colder water. In other words, the colder the water, the more oxygen the water can hold.

Water is 800 times denser than air and contains 95 percent less oxygen. Fish expend a considerable amount of energy breathing but are able to extract 80 percent of the available oxygen from water, as opposed to humans, who extract only 25 percent of the oxygen they breathe in from the atmosphere. The saturation level of oxygen in water is in the region of seven to eight parts per million. Under normal conditions, there is never too much oxygen in a fish tank. Such only happens when an operator purposely attempts to super saturate the water. Operators can super saturate the water by diffusing 100 percent liquid oxygen into the water, but there is no real value to doing so.

Too much oxygen is also harmful to fish. Hyperoxia is the state of water when it holds an extremely high amount of oxygen. At this state, water is described as having a dissolved oxygen saturation of greater than 100 percent. This percent can be 140-300 percent. If fish are exposed to such water, their blood equilibrates, bubbles form in the blood,

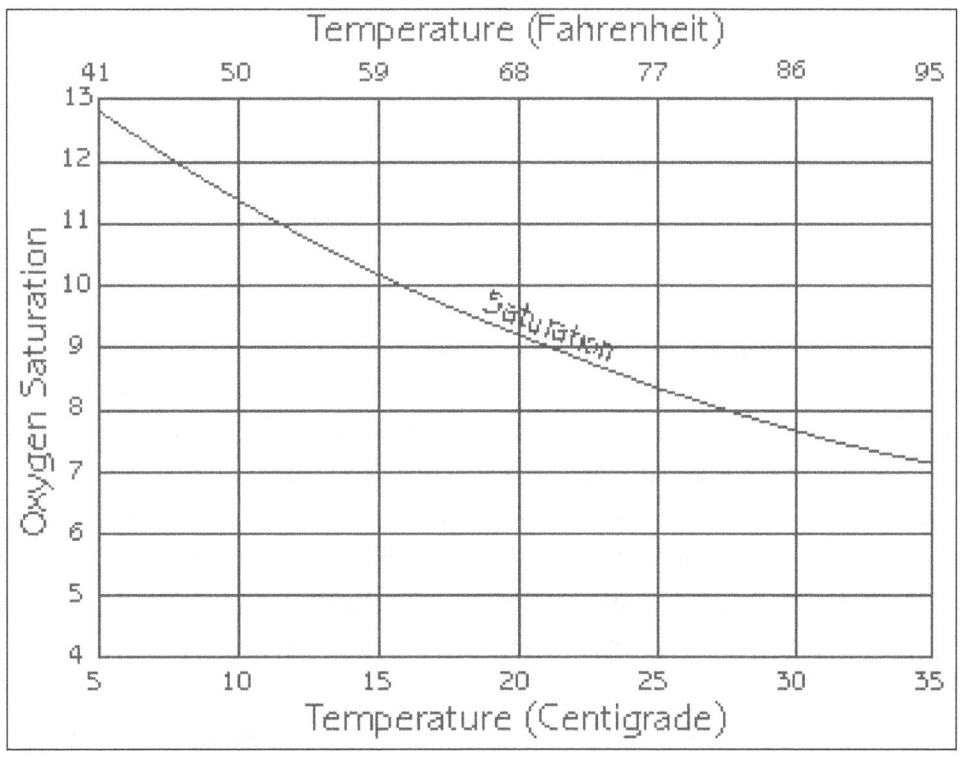

FIGURE 77. Oxygen and temperature relationship.

TABLE 24.

BENCHMARK	OXYGEN DEFICIENCY	ALGAE BLOOM	DISEASE
Fish Behavior	Swimming Near Surface Gasping for Air	Abnormal Swimming	Abonorma Swimming
Time of Fish Kill	Nighttime into Morning	Brightest Part of Day	Anytime
Size of Dead Fish	Large Fish Die First	Small Fish Die First	Small Fish Die First
Microscopic Algae Abundance	Algae Dying	One Dominant Algal Species	No Effect
Dissolved Oxygen Concentration	Less than 3 ppm Oxygen	Supersaturated Oxygen Concentrations	No Effect
Water Color	Brown or Gray or Black	Dark Green or Brown or Golden	No Effect

and can block the capillaries. In severe cases, death occurs rapidly as a result of blockage of the major arteries. The remedy is either to remove the fish to normally equilibrated water or to provide vigorous aeration to strip out the excess gas.

Most fish, including **Tilapia, prefer dissolved oxygen levels above 5 ppm.** Ideally, dissolved oxygen levels will never drop below 3 ppm.

The fish will suffocate from low DO. Symptoms of low DO include fish swimming close to the water surface grasping for air, fish lying on the bottom of the tank, and red gills. However, waiting till then is not a great idea. Preventative measure should always be taken prior to the health of your fish being jeopardized. Therefore, be sure you have operational features built into your system so that there is always adequate dissolved oxygen present. If your system does not naturally generate enough DO, then simply adding an aeration device or air pump will ensure that there is sufficient supply for the fish to thrive.

To measure DO, use a DO meter. This is typically the most expensive piece of measuring equipment used in aquaculture. Although many operators get by without them, they do so at a risk. If you have the money to spend, they are nice to have on hand. "Dissolved oxygen meters range in price from about $150 to $500 (USD, 2021).

It is important to have a backup aerator (battery or generator operated) that is independent of the main power supply. This ensures that air can be continually pumped into the water to keep the fish alive, in case of a power failure.

As mentioned previously, temperature has an especially important effect to dissolved oxygen. As the temperature of the water goes up, the water loses the ability to hold the dissolved oxygen and the concentration goes down. When the water cools, it regains the ability to hold higher amounts of oxygen. Knowing this relationship, one can deduce that hypoxia tends to occur in the warmer months of the year, namely during the summer.

Aeration Sizing

The aeration rate is measured in cubic feet per minute (CFM). CFM is the measurement of the volume of air flow beingintroduced into the system.

Air pressure is the pressure required to deliver the correct amount of air flow for proper aeration. Air pressure is measured in terms of pounds per square inch. **Professional aquaponic and aquaculture operators strive to ensure that 1 cfm per 300 gallons is being achieved.** This is typically not a problem when the system is being properly circulated. However, aeration devices, oxygen injectors, or oxygen diffusers are necessary if the fish tank does not have adequate circulation (dead spaces), and/or there are high stocking densities.

As figure 78 illustrates below, dissolved oxygen increases throughout the day and decreases throughout the night. The most critical time to aerate is just before dawn.

Pump Size

The pump(s) should cycle the total volume of fish tank water once every two hours, but ideally once every hour. If the pump is on a timer, then the 'on' and 'off' phases need to be considered, so that the tank volume can still be pumped within an hour.

Clean Water

Don't be fooled by looks. Clear water is not always necessarily clean. Harmful substances such as nitrite, carbon dioxide, as well as other pollutants, do not discolor water. Low DO levels also do not necessarily alter water color. On the other hand, green water, while not clear can actually be healthy for fish, particularly Tilapia. Therefore, monitoring water quality on a regular basis is critical.

Chlorine

Chlorine kills fish and is also harmful to humans. If water must be added directly to the tank, it is best to use non-chlorinated well water or run water through a filter that removes chlorine prior to adding it to tank. Otherwise set up a temporary holding tank, such as a 55-gallon drum, where the chlorinated water can be stored for 24 hours in advance of adding it to the aquaculture system. This will provide an opportunity for the chlorine to dissipate from the water before becoming part of the system (this process is also referred to as 'off gas').

It is prudent to keep spare dechlorinated water readily on hand. It is recommended that you store at least 10 percent of your system water volume in a separate container. If using municipal water, be sure to let it sit for a minimum of 24 hours before adding it to your system so it can properly dechlorinate. Water will need to be added to the system on a regular basis to offset evaporation.

Light

Fish are sensitive to light, and some species much more than others (e.g., largemouth bass). Avoid any direct lighting to the fish tank. Even a fish tank positioned under a typical room light can stress the fish. Therefore, it is prudent to ensure that the only lighting of the fish tank, or fish tank area, be that of the

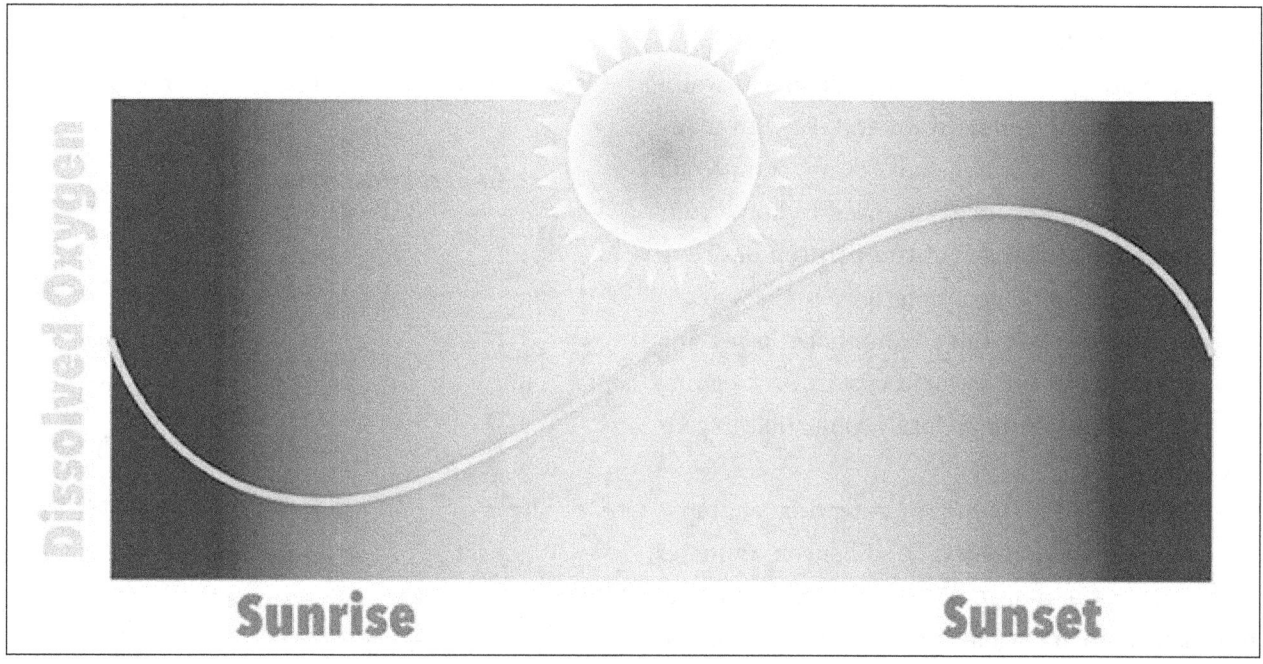

FIGURE 78. The normal daily cycle of dissolved oxygen production.

ambient lighting type. Ambient lighting is a general illumination that comes from all directions in a room that has no visible source. This type of lighting is in contrast to directional lighting. Even so, the ambient light should not be too bright.

Water Hardness

Water hardness is a part of the aquaculture water chemistry that is often not fully understood. However, don't let the subject intimidate you. It is not that complicated. Fish tank water hardness is measured in degrees of hardness. Many home aquarium water test kits will give you measurements in either degrees of hardness (dh) or in parts per million (ppm).

When we discuss water hardness, we are simply looking at the amount of dissolved minerals in our water. There are two distinct measurements of water hardness. When you test your water, you will test for general hard-ness (GH) and carbonate hardness (KH).

General Hardness (GH)

The general hardness (GH) of your aquaculture water is a measurement of dissolved magnesium and calcium. The (GH) of aquaculture water can have great effects on your fish, so it is particularly important to ensure that the fish you choose to keep will thrive in your water or adjust the water accordingly.

There are species of fish that live in soft water and others that live in hard water. Like with many other aspects of raising fish, it is important to do your research in advance. You need to know your water quality, pretenses of the fish you desire to raise, and what would need to be done to the water to accommodate the fish. This will minimize losses. You can also keep water treatment cost down by going with a species that is most closely compatible to your water supply.

Most fish will survive in the water from our tap -- after chlorine, chloramines, and other contaminates are removed -- with no problems. Unless tests reveal unusually hard or extremely soft water, there should not be a problem. Most fish will adapt fine. However, successful breeding and the overall coloration of the fish can be directly linked to the water hardness.

Carbonate Hardness (KH)

The second component of water hardness is the carbonate hardness (KH). (KH) is the measurement of carbonate and bicarbonate ions in your water. In simple terms, the (KH) of water is a measurement of the buffering capacity of your water. The (KH) of water will determine how much your pH will fluctuate. The higher the (KH) is in your water, the more stable your pH will remain. When (KH) drops, so too will your pH. There is a distinct relationship between (KH) and pH. If you have ongoing problems with the pH dropping, check the (KH) levels in the tank water.

How to Soften Aquaponics Water

There are several ways to soften your aquaculture water if your tap water is too hard for your fish. However, if you do need to soften your aquaculture water, do so slowly. Any drastic changes can shock the fish, cause injury, or even result in death.

Reverse Osmosis

The most economical and popular method to soften your aquaculture water is to use a Reverse Osmosis (RO) system. These units remove heavy

TABLE 25. **General Hardness Chart**

DEGREE OF HARDNESS (DH)	ppm	HARDNESS
0–3	0–50	Very Soft
6–Mar	51–100	Soft
12–Jun	101–200	Slightly Hard
18–Dec	201–300	Moderately Hard
18–30	301–450	Hard
30+	450+	Very Hard

metals, minerals, and contaminants from your water source.

Water Softening Pillows

Water softening pillows can be used on small systems. There are several different manufactures of water softening pillows. These pillows work well for aquaculture systems that are less than 50 gallons. You simply channel the water through the pillow material. It softens the water through ion-exchange. Many pillow softeners can be recharged and reused by soaking the pillow in a saltwater solution. It is then placed in the tank where the sodium ions are released into the water and replaced by calcium and magnesium ions. After a few hours or days, the pillow (along with the calcium and magnesium) is removed, and the pillow recharged again.

How to Increase Aquaculture Water Hardness

If you find that your water source is too soft, you may need to harden your aquaculture water. There are several things you can do to harden your aquaculture water.

Crushed Coral

Adding crushed coral to your tank can help increase the water hardness. Some operators burdened with a hard water sources, choose to use crushed coral as their substrate media or integrate it into other media types creating a media mix that resolves the problem.

Limestone

Using limestone allows for calcium and other minerals to leach out into the water column. The most popular limestone used in cichlid tanks is Texas Holey Rock Limestone.

Buffer Additives

There are several buffers on the market that will help raise GH and KH while maintaining the pH of the water. Most use a combination of different salts including carbonate salts to increase water hardness. Be sure to closely follow the directions on the manufacture's label.

Aquaculture Water Hardness Tips

- Maintaining a regular schedule of water changes to ensure high water quality will typically prevent your pH, GH, and KH from swinging back and forth so much.
- If you decide to dose additives, be sure they are safe to enter your food chain and follow the directions on the manufacturer's label.
- When you use limestone or crushed coral, ensure that you clean it thoroughly.
- When adjusting any water parameters such as aquaculture water hardness, adjust slowly. You can kill your fish with extreme changes.
- Purchase a high-quality water test kit.

Algae

Photosynthetic growth and activity by algae in aquaculture and aquaponic systems affect the water quality parameters of pH, DO, and nitrogen levels. Algae are a class of photosynthetic organisms that will readily grow in any body of water that is rich in nutrients and exposed to sunlight. Some algae are microscopic, single-celled organisms called phytoplankton, which can color the water green. Macroalgae are much larger, commonly forming filamentous mats attached to the bottoms and sides of tanks.

It is important to prevent algae growing in an aquaculture and aquaponic systems because they are problematic for several reasons. Algae act as both a source and sink of DO, producing oxygen during the day through photosynthesis and consuming oxygen at night during respiration. They can dramatically reduce the DO levels in water at night to the point of even being fatal to fish. This production and consumption of oxygen is related to the production and consumption of carbon dioxide, which causes daily shifts in pH as carbonic acid is either removed (daytime — higher water pH) from or returned (nighttime — lower water pH) to the system. Finally, filamentous algae can clog drains and block filters within the unit, leading to problems with water circulation. It should be noted that some aquaculture operations benefit greatly from culturing algae for feed, referred to as green-water culture, including tilapia breeding tanks, shrimp culture, and biodiesel production.

Algae Control

Algae are unsightly and, on drying, it can emit a foul odor. As mentioned above they will also rob the water of minute elements, including oxygen. Algae are not harmful to plants, except on rare occasions where there is stagnant water when it can harbor insects and diseases. Some people will even argue that algae give off certain enzymes which are beneficial to plants, and they openly encourage its growth. The main problem with algae is that it pulls oxygen outof the water and, as a result, can be harmful and even deadly to fish.

Again, algae require two things to flourish: light and oxygen. If one or the other is not present, algae will not grow. In aquaculture, algae growth can be limited or excluded by minimizing light. Do not let any direct light shine on the fish tank (only ambient lighting).

Fish tanks should be shaded. Outdoor fishponds can have a shade cover or D.W.C. raft(s) floating on the surface of the water to provide shade.

Nitrite

Nitrite enters a fish culture system after fish digest feed and the excess nitrogen is converted into ammonia, which is then excreted as waste into the water. Total ammonia nitrogen (TAN; NH_3 and NH_4^+) is then converted to nitrite (NO_2) which, under normal conditions, is quickly converted to non-toxic nitrate (NO_3) by naturally occurring bacteria. Uneaten (wasted) feed and other organic material also break down into ammonia, nitrite, and nitrate in a similar manner. Brown blood disease occurs in fish when water contains high nitrite concentrations. Nitrite enters the bloodstream through the gills and turns the blood to a chocolate-brown color. Hemoglobin, which transports oxygen in the blood, combines with nitrite to form methemoglobin, which is incapable of oxygen transport. Brown blood cannot carry sufficient amounts of oxygen, and affected fish can suffocate despite adequate oxygen concentration in the water. This accounts for the gasping behavior often observed in fish with brown blood disease, even when oxygen levels are relatively high.

Nitrite problems are typically more likely in closed, intensive culture systems due to insufficient, inefficient, or malfunctioning filtration systems. High nitrite concentrations in ponds occur more frequently in the fall and spring when temperatures are fluctuating, resulting in the breakdown of the nitrogen cycle due to decreased plankton and/or bacterial activity.

A reduction in plankton activity in ponds (due to lower temperatures, nutrient depletion, cloudy weather, herbicide treatments, etc.) can result in less ammonia assimilated by the algae, thus increasing the load on the nitrifying bacteria. If nitrite levels exceed that which resident bacteria can rapidly convert to nitrate, a buildup of nitrite occurs, and brown blood disease is a risk. Although nitrite is seldom a problem in systems with high water exchange rates or good filtration, systems should be monitored year-round and managed, when necessary, to prevent severe economic loss from brown blood in any fish culture facility.

Nitrite — Susceptibility of Fish Species to Brown Blood Disease

Largemouth and smallmouth bass, as well as bluegill and green sunfish, are resistant to high nitrite concentrations. Catfish and **tilapia are sensitive to nitrite**, while trout and other cool water fish are sensitive to extremely small amounts of nitrite. Goldfish and fathead minnows fall in between catfish and bass in their susceptibility to brown blood disease resulting from high nitrite levels. Striped bass and its hybrids appear sensitive to nitrite, but little is known about the relative sensitivity compared to other species.

Fluoride

Water fluoridation is the practice of adding industrial-grade fluoride chemicals to water for the purpose of preventing tooth decay. The fluoride being added to public drinking water is actually an industrial waste product. One of the little-known facts about this practice is that the United States, which fluoridates over 70 percent of its water supplies, has more people drinking fluoridated water than the rest of the world combined. Most developed nations, including all of Japan and 97 percent of western Europe, do not fluoridate their water.

Contrary to popular belief, comprehensive data from many unbiased scientific studies (as well as the World Health Organization) have proven that there is no discernible difference in tooth decay between the minority of western nations that fluoridate water, and the majority of nations who don't fluoridate their water. In fact, the tooth decay rates in many non-fluoridated countries are lower than the tooth decay rates in fluoridated ones.

Fluoride is extremely toxic to fish, plants, and humans. Check out the warning label on the back of your toothpaste, then immediately throw it in the trash, and start using non-fluoridated toothpaste from now on.

Harmful Effects of Fluoride

Most fluoride that is added to municipal water is an unnatural form of fluoride that contains sodium. It is over 80 times more toxic than naturally occurring calcium fluoride.

The fluoride ion (F-) is extremely reactive and strongly attracted to calcium. Its preference for calcium overrides its attraction to other ions. In nature, fluoride is most often bound to calcium. When sodium fluoride is ingested, it rapidly robs the body of calcium. In fact, sodium fluoride poisoning results when calcium is stolen from the blood.

Fluoride has the ability to affect other chemicals and heavy metals, in some cases making them even more harmful than they would be on their own. For example, when you combine chloramines with the hydrofluorosilicic acid added to the water supply, they become highly effective at extracting lead from old plumbing systems, promoting the accumulation of lead in the water supply.

Studies have shown that hydrofluorosilicic acid increases lead accumulation in bone, teeth, and other calcium-rich tissues. This is because the free fluoride ion acts as a transport of heavy metals, allowing them to enter into areas of your body they normally would not be able to go, such as into your brain.

Fluoride has long been known to be a very toxic substance. This is why, like arsenic, fluoride has been used in pesticides and rodenticides (to kill rats, insects, etc.). It is also why the Food and Drug Administration (FDA) now requires that all fluoride toothpaste sold in the U.S. carry a poison warning that instructs users to contact the poison control center if they swallow more than used for brushing.

Excessive fluoride exposure is well known to cause a painful bone disease (skeletal fluorosis), as well as a discoloration of the teeth known as dental fluorosis. Excessive fluoride exposure has also been linked to a range of other chronic ailments including arthritis, bone fragility, dental fluorosis, glucose intolerance, gastrointestinal distress, thyroid disease, possibly cardiovascular disease, and certain types of cancer.

Certain subsets of the population are particularly vulnerable to fluoride's toxicity. Populations that have heightened susceptibility to fluoride include infants, individuals with kidney disease, individuals with nutrient deficiencies (particularly calcium and iodine), and individuals with medical conditions that cause excessive thirst.

For a complete breakdown of all the harmful effects of fluoride, please refer to the *Fluoride Action Network* (FAN). FAN's work has been cited by national and international media outlets including the *New York Times*, *Wall Street Journal*, *TIME Magazine*, *National Public Radio*, *Scientific American*, and *Prevention Magazine* among others. *FAN* is an official project of the *American Environmental Health Studies Project (AEHSP)* — a registered non-profit (501c3) organization. FAN's website address is: http://fluoridealert.org/faq

Fluoride negatively impacts health, and it is prudent to avoid it. The following is a brief list of some of the detrimental health consequences of ingesting fluoride:

- Gastrointestinal Effects
- Bone Fractures
- Brain Effects
- Cancer
- Cardiovascular Disease
- Diabetes
- Endocrine Disruption
- Acute Toxicity
- Hypersensitivity
- Kidney Disease
- Male infertility
- Pineal Gland
- Skeletal Fluorosis
- Thyroid Disease

Fluoride Side Note

We all need to beware of other ways we are constantly exposed to fluoride. Not only is one source bad enough (e.g., drinking water, toothpaste, etc.), but the accumulation from multiple sources is especially harmful.

One of the primary sources of fluoride exposure is non-organic foods, due to the high amounts of fluoride-based pesticide residues on these foods. Non-organic foods may account for as much as one-third of the average person's fluoride exposure. Foods particularly high in fluoride include non-organic fresh produce, breakfast cereals, juices (particularly grape juice), deboned meats such as lunch meats, other meats through food chain accumulation, and black or green tea (even if organic). Unfortunately, food labels don't address the quantity of fluoride in food.

How to Remove Fluoride from Water

Unfortunately, water filtration does not remove fluoride. Many water filters (e.g., Brita & Pur) use an "activated carbon" filter that does not remove fluoride. The fluoride molecule is smaller than the water molecule, therefore it cannot be removed by filtration.

If you live in a community that fluoridates its water supply, there are several options to avoid drinking the fluoride that is added. Unfortunately, each of these options will cost money (unless you happen to have access to a free source of spring water). These options include:

- **Spring water:** Most spring water contains very low levels of fluoride (generally less than 0.1 ppm).
- **Reverse Osmosis:** Reverse Osmosis can remove between 90 and 95 percent of fluoride. Contaminants are trapped by the RO membrane and flushed away in the wastewater. The process requires between two and four gallons of water to produce one gallon of RO water (depending on the quality of the water and the efficiency of the RO unit).

- **Water Distillation**: Distillation is capable of removing just about anything (except volatile compounds) from water. Distilling water is an effective way of removing fluoride from water. The drawback to distillation is that the processsis time and energy consumptive. Distillation also leaves the resulting water empty and lifeless. If you use distilled water, you need to do research on how to add minerals back to the water following the distillation process, often referred to as the 'full-spectrum living water'.

Sources of Water for Aquaculture

On average, an aquaculture system uses 1–3 percent of its total water volume per day, depending on the location and environmental factors where the fish are being raised. Water from direct evaporation and splashing. As such, this waterloss will need to be replenished periodically. The water source used will have an impact on the water chemistry of the unit. This section will review some common water sources and the common chemical composition of that water. New water sources should always be tested for pH, hardness, salinity, chlorine, and for any pollutants in order to ensure the water is safe to use.

Salinity

Now is a good place to discuss a water parameter not yet addressed in detail — salinity. Salinity indicates the concentration of salts in water, which include table salt (sodium chloride — NaCl), as well as plant nutrients, which are in factsalts. Water salinity can be measured with an electrical conductivity (EC) meter, a total dissolved solids (TDS) meter, a refractometer, or a hydrometer or operators can refer to local government reports on water quality.

Salinity is measured either as conductivity, or how much electricity will pass through the water, in units of microsiemens per centimeter (µS/cm), or in TDS as parts per thousand (ppt) or parts per million (ppm or mg/liter). For reference, seawater has a conductivity of 50,000 µS/ cm and TDS of 35 ppt (35 000 ppm). It is recommended that low salinity water sources be used. Salinity, generally, is too high if sourcing water has a conductivity more than 1 500 µS/cm or a TDS concentration of more than 800 ppm. Although EC and TDS meters are commonly used for hydroponics to measure the total amount of nutrient salts in the water, these meters do not provide a precise reading of the nitrate levels, whichcan be better monitored with nitrogen test kits.

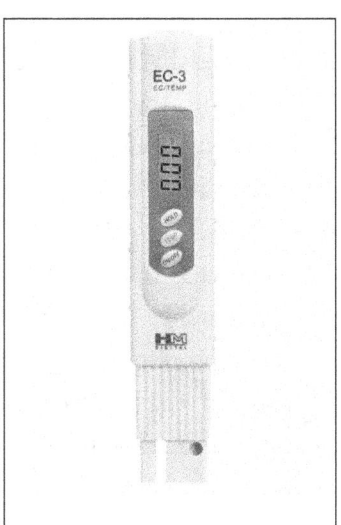 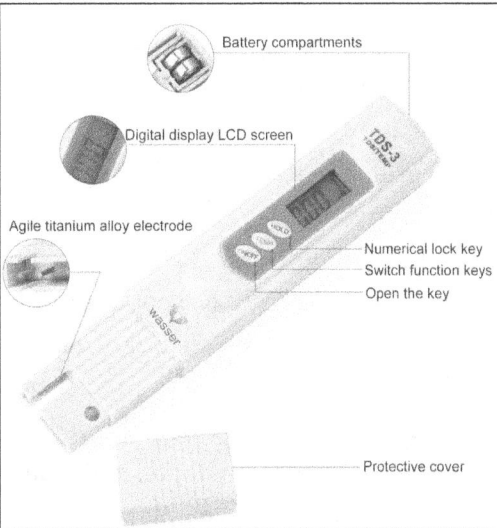 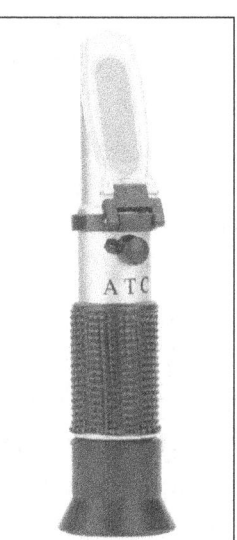 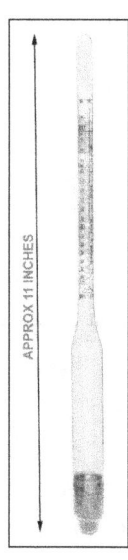

FIGURE 79. (Left to Right): EC meter, TDS meter, Refractometer, Hydrometer

NOTE: All above meters, or an equal, can be acquired for less than $25 each. Several of these meters measure multiple water constituents.

FIGURE 80. *Rainwater collection examples*

Rainwater

Collected rainwater is an excellent source of water. The water will usually have a neutral pH and exceptionally low concentrations of both types of hardness (KH and GH) and almost zero salinity, which is optimal to replenish the system and avoid long-term salinity buildups. However, in some areas affected by acid rain, as recorded in a number of localities in eastern Europe, eastern United States of America, and areas of southeast Asia, rainwater will have an acidic pH. If this is the case in your location, it is good practice to buffer rainwater and increase the KH.

When collecting rainwater, make sure that the gutters and roof do not contain any chemicals that may leach into your water from roofing materials. It is prudent to test the collected rainwater in the beginning to ensure that it does not contain contaminants. Also, depending upon your location, you made need to beware of the potential for harmful concentrations of air pollution residue on the roof, which would also negatively impact your system. Lastly, in some jurisdictions it is illegal to catch rainwater. These government jurisdictions have decided that homeowners don't own the rain that falls on their property.

A well-designed system includes an overflow pipe to protect against damage caused by the tank overflowing during periods of heavy rains or low usage. All overflow water should be discharged away from foundations and other structures.

- One inch of rain falling on a square foot of surface yields approximately 0.6 gallons of water. Using the following equation, it is easy to determine how much rainwater can be collected.

- Rain caught (gallons) = (inches of rain) x 0.6 x (portion of building footprint)

For example, if your home's footprint is 1,400 square feet and you want to know the amount of water that comes from a ¼ inch (.25") rain event:

- Rain caught (gallons) = (.25) x (.6) x (1,400) = 210 gallons (or less if you're only gathering from one part of the roof).

Rainwater collection can be easily achieved by connecting a large clean container to water drainage pipes surrounding a building or house (see figure 80).

Collecting rainwater is relatively easy but storing rainwater can be a bit more challenging. The water has to be retained until needed, and the water has to be kept clean. The storage container(s) should be covered with a screen to prevent mosquitoes and plant debris from entering. Depending on the intended uses of your collected rainwater, some form of filtration and/or disinfection of the water that comes from the storage tank may be necessary. Do your research and take the necessary protection measures.

Cistern or Aquifer Water

The quality of water taken from wells or cisterns will largely depend on the material of the cistern and bedrock of the aquifer. If the bedrock is limestone, then the water will probably have quite high concentrations of hardness, which may have an impact on the pH of the water. Water hardness is not a major problem in aquaponics, because the alkalinity is naturally consumed by the nitric acid produced by the nitrifying bacteria. However, if the hardness levels are remarkably high it may be necessary to use small amounts of acid to reduce the alkalinity before adding the water to the system in order to prevent pH swings within the system. Aquifers on coral islands often have saltwater intrusion into the fresh- water lens, and can have salinity levels too high for aquaculture, so monitoring is necessary, and rainwater collection or reverse osmosis filtration may be better options.

Tap or Municipal Water

Water from the municipal supplies is often treated with different chemicals to remove pathogens. The most common chemicals used for water treatment are chlorine and chloramines. These chemicals are toxic to fish, plants, and beneficial bacteria. They are used to kill bacteria in water and as, such are detrimental to the health of the fish. Chlorine test kits are available, and if high levels of chlorine are detected, the water needs to be treated before being used.

The simplest method is to store the water before use, thereby allowing all the chlorine to dissipate into the atmosphere. This can take upwards of 24 hours but allowing 48 hours is even better. Dissipation of chlorine will occur faster if the water is heavily aerated with air stones. Chloramines are more stable and do not off-gas as readily. If the municipality uses chloramines, it may be necessary to use chemical treatment techniques such as charcoal filtration or other dechlorinating chemicals. Even so, off-gassing is usually enough in small-scale units using municipal water. A good guideline is to never replace more than 10 percent of the water without testing and removing the chlorine first. Moreover, the quality of the water will depend on the water source and treatment methods being implemented by the public water provider. Always check new sources of water for hardness levels and pH and use acid if appropriate and necessary to maintain the pH within the optimum levels indicated above.

Filtered Water

Depending on the type of filtration (i.e., reverse osmosis or carbon filtering), filtered water will have most of the metals and ions removed, making the water very safe to use and relatively easy to manipulate. However, like rainwater, deionized water from reverse osmosis will have low hardness levels and may need to be buffered.

Water Testing

In order to maintain good water quality, it is prudent to perform water tests once per week to make sure all the parameters are within the optimum levels. Seasoned aquaculture professionals do not need to test the water as often as newbies. They typically only test if a problem is suspected. However, daily health monitoring of the fish growing in the unit will indicate if something is wrong, although this method should not be a substitution for testing the water.

Color-coded freshwater test kits are readily available and easy to use. These kits include tests for pH, ammonia, nitrite, nitrate, GH, and KH. Each test involves adding five to ten drops of a reagent into 5 milliliters of aquaculture water; each test takes no more than five minutes to complete. Other methods include some of the meters and water testing tools referenced in the sections above or water test strips, which are inexpensive and moderately accurate (see figure 82).

The most important tests to perform weekly are pH, nitrate, carbonate hardness and water temperature, because these results will indicate whether the system is in balance. The results should

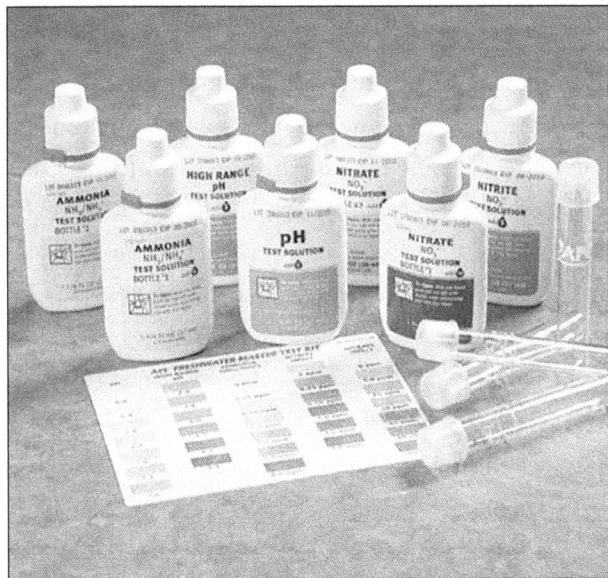

FIGURE 81. Rainwater Collection

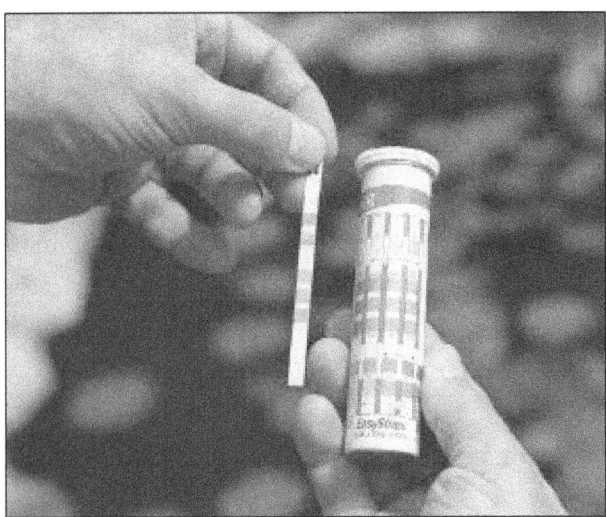

FIGURE 82. Water test strip kit.

be recorded each week in a dedicated logbook so trends and changes can be monitored throughout growing seasons. Testing for ammonia and nitrite is also extremely helpful in order to diagnose problems in the system, especially in new systems or when significant changes occur, such as with a major fish harvest, or if there is increase in fish mortality raises toxicity concerns in an ongoing system. Although with an aquaponic system, weekly monitoring may not be necessary in established systems, the testing can provide extraordinarily strong indicators of how well the bacteria are converting the fish waste and provide feedback as the overall health of the system. Testing for ammonia and nitrate should be the first priority if any problems are observed with the fish or plants (aquaponics).

Introducing Fish into the System

Ensure the pH is at the desired level for fish. A pH of 7 to 9 is ideal for Tilapia. Ensure the fish tank water temperature is within the recommended range for your fish.

Feeding Fish

A good rule of thumb is to feed your fish as much as they will eat in 5 minutes, 1 to 3 times per day. An adult fish will eat approximately 1 percent of its bodyweight per day. Fish fry (babies) will eat as much as 7 percent. Be sure your fish are being fed enough. However, be cognizant of the fact that over feeding fish will negatively affect water quality, is wasteful, and is an unnecessary increase in cost.

Fish not eating as they should, is a good indication they are stressed or unhealthy. Some factors that may result in fish not eating as they should include

- Living in conditions outside of their optimal temperature range
- Water quality issues: Improper pH range, too much ammonia in the system, inadequate dissolved oxygen
- Loud or irritating noises and vibrations
- Direct lighting upon the fish tank

Starting the System

- Start out by adding only 1/2 as many fish as it would take for the fish tank to be fully stocked.
- Test daily for elevated ammonia and nitrite levels. If either becomes too high perform a partial water exchange.
- During the start-up cycling period, feeding the fish only once a day will help control ammonia levels.

System Maintenance

Ammonia and Nitrite levels should be less than 0.75 ppm. If ammonia levels rise suddenly, check to see if there is a dead fish in your tank. If Nitrite levels rise undesirably, something has likely occurred that has damaged the bacteria environment in the system. If either of these circumstances occur, stop feeding the fish until the levels stabilize, and, in extreme cases, do a 1/3 water exchange to dilute the existing solution. Nitrates can rise as high as 150 ppm without causing problems. If nitrates exceed 150 ppm, it would be prudent to harvest some fish, add additional plants or expand the system by adding another grow bed.

Summary

- Water is the lifeblood of an aquaculture system. It is the medium through which the fish receive their oxygen. It is especially important to understand water quality and basic water chemistry in order to properly manage aquaculture.
- There are five key water quality parameters for aquaponics: dissolved oxygen (DO), pH, water temperature, total nitrogen concentrations and hardness (KH). Knowing the effects of each parameter on fish is essential.
- There are simple ways to adjust pH. Bases, and less often acids, can be added in small amounts to the water in orderto increase or lower the pH, respectively. Acids and bases should always be added slowly, deliberately, and carefully. Rainwater can be alternatively used to let the system naturally lower the pH through nitrifying bacteria consumingthe system's alkalinity. Calcium carbonate from limestone, seashells, or eggshells increases KH and buffers pH against the natural acidification.
- Some aspects of the water quality and water chemistry knowledge needed for aquaculture can be complicated,in particular the relationship between pH and hardness, but basic water tests are used to simplify water quality management.
- Water testing is essential to maintaining good water quality in the system. Test and record the following water quality parameters each week: pH, water temperature, nitrate and carbonate hardness. Ammonia and nitrite tests should be used especially at system start-up and if abnormal fish mortality raises toxicity concerns.
- It is important that the beginner not be intimidated or overwhelmed by the water testing and water quality managementprocess. Everything discussed in this chapter can be easily learned. Furthermore, experience is an excellent teacher.
- A pH of 7 to 9 is ideal for Tilapia.
- Most aquaculture fish, including Tilapia, prefer dissolved oxygen levels above 5 ppm.

PART IV : OPERATION AND MAINTENANCE

CHAPTER 16

Fish Breeding, Fish Reproduction, and Raising Your Own Crop of Fish

Breeding and Raising Fish Overview

Breeding fish successfully requires knowledge, effort, and attention to details, but it also provides many rewards. Besides the personal gratification one acquires through the process, a great deal of money can be saved, and gained, as a result. The increasing demand for fish and fish protein has resulted in widespread overfishing, increased prices, and diminishing supply. Throw in the fact that the general public is becoming more knowledgeable about our fishfood supply being heavily contaminated with heavy metals (including mercury) and radioactive isotopes, it makes raising your own fish a very lucrative endeavor. Also, replenishing your aquaculture fish tank with your own bred fish, rather incurring the cost of another supply of fingerings (as well as taxes, shipping, and handling fees) can greatly reduce your operational expenses.

It is difficult to raise fish from young in the main fish tank. Using separate tanks for breeding and raising the fry is the best approach. The more conditions that are right within the tank, the greater your fish breeding success rate, although some species, such as tilapia, are quite forgiving and make fish breeding a relatively easy process.

Many fish will spawn if you place a male and female in their own tank and give them a green "spawning mop" or create a dark "pot cave". To make a pot cave, simply place a terracotta pot on its side, and then fill it approximately ¼ full of sand. This setup will work for many different species. Most will use the sand, some will use the hard, under-surface, while others will use the crevice created on the outside of the pot. Some fish require a certain speed of water current. Temperature, pH, and overall water quality are also important. For some species, the male and/or mother will need to be removed at some point in the process. Generally, fry can be fed small pieces of flakes, brine shrimp, small worms, and soaked oatmeal. Do your research and know your fish species to understand their natural habitats for determining which methods and food options best suit them.

There are already many resources readily available which provide detailed, systematic instructions on breeding and raising all common aquaculture friendly fish. This book will not attempt to address all the specifics of breeding for each species. Rather, this chapter will describe what it takes to successfully breed tilapia. Although most of what is being described in this chapter will also apply to other species, it would be prudent for you to investigate the reproduction needs of your fish species if you are not raising tilapia.

Breeding Tilapia

Tilapia are classified as either mouth brooders or substrate spawners. All tilapias are prolific breeders. With the proper environmental conditions, tilapia can easily reproduce and provide an ample fish supply for consumption and commercial success.

Mouth Brooders

Members of the Oreochromis genus are maternal mouth brooders and are a common choice for aquaponics or aquaculture. In terms of popularity, the Nile Tilapia (*O. niloticus*) is the most widely cultured tilapia, followed by Blue Tilapia (*O. aureus*), and Mozambique Tilapia (*O. mossambicus*).

The *Oreochromis* display an elaborate courtship behavior. After building a nest, the male aggressively repels other males that enter into proximity of the nest. When ready to spawn, the male displays a darkened color and leads a female to the nesting area. The fish then swim around the nest and the male will butt against the female genital area to induce egg laying. The courtship is often brief, lasting only a few minutes in many cases and seldom more than a few hours.

The female tilapia lay their eggs in pits (nests) and after fertilization by males. The female collects the eggs in her mouth (buccal cavity) to maintain them until hatching.

Other tilapia display different mouth brooding behavior. *Sarotherodon galilaeus* are bi-parental, with both parents brooding the eggs and defending the newly hatched fish. The male *Sarotherodon melanotheron* is the parent that performs the mouth brooding, while the female leaves the nest.

Substrate Spawners

Tilapia rendalli and *tilapia zillii* are two popular, commercially raised species that are substrate spawners. The male and females will build a nest and defend it together. A male and female will typically form a bonded-mating-pair and courtship can last up to a week, but usually takes place over several days.

Females will first lay their eggs in pits (nesting area) dug in the bottom of a lake or pond. You can simulate this condition in a tank by adding some substrate (e.g., gravel) which allows the tilapia to evacuate a nesting area. The male will then spawn and fertilize the eggs. After fertilization, the parents guard the eggs, chasing away predators and making sure proper aeration is maintained for hatching.

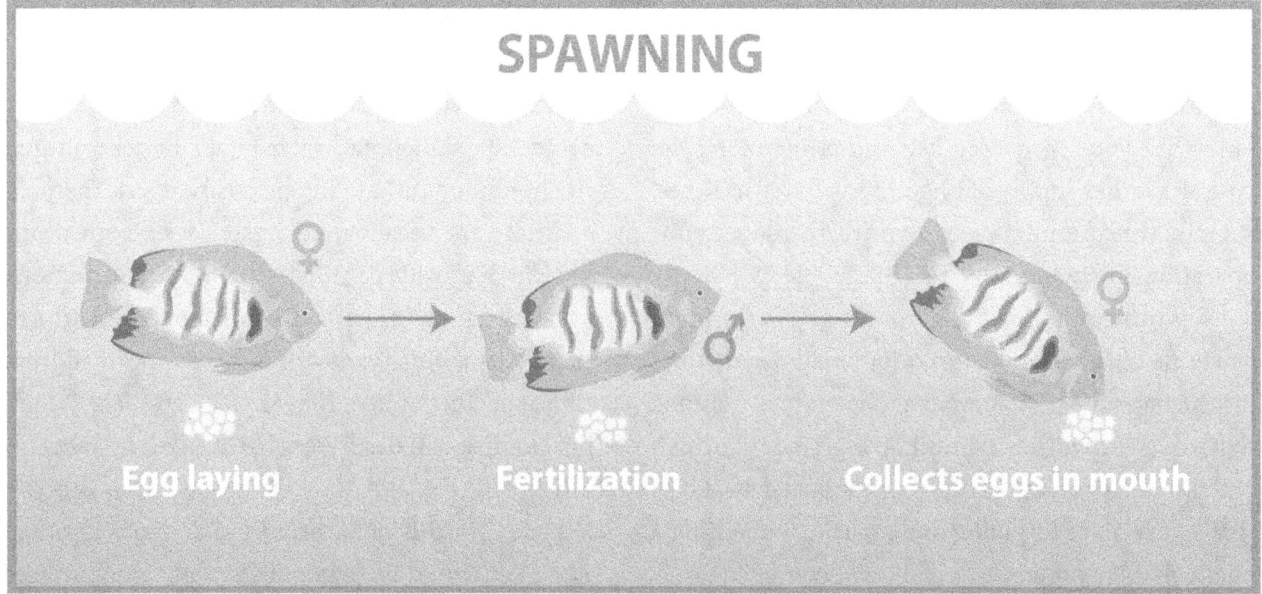

FIGURE 83. Tilapia spawning process.

Frequent Breeding and Mouth Brooding

With the proper set up, at temperatures of 85°F (29°C), they can produce baby tilapia (fry) almost every week, year-round. The mouth brooding and maternal protection of the fry helps to create a high survival rate. This combination of continuous production and high survival rate allows the operator to have a constant supply of fingerlings to replace those that get big enough to eat.

Tilapia Fish Farming Stages

The process for farming Tilapia includes the following stages:

Breeding → Fry sizing → Fingerling production « Grow-out to plate/market size → Purging → Harvesting → Processing → Packaging « Marketing → Cooking → Eating

Tilapia Breeding Fundamentals

Tilapia will reproduce profusely if adults are well fed, and the young can find refuge. If hungry, the adults will cannibalize their young to some degree, but rarely will they control their own population. They prefer many other food items instead of their own young. However, non-spawning adults do have a seemingly insatiable appetite for eggs. Juveniles from previous spawns will actually be the most cannibalistic fish in the tank. They will eat any sibling they can fit in their mouth.

One female will typically produce about 200-1,000 eggs per spawn, and she will spawn every four to five weeks or so if conditions are right. Even if there is a low survival rate, that is still a lot of tilapia recruitment. In an average small system, a single female could have a tank filled to the brim with young tilapia in a relatively short period.

In large-scale tilapia farming operations, only the male population is raised, to avoid wild-spawning and eliminate the small size of females. These commercial growers use hormones to convert female fry to male fry. This practice is rarely done in small to mid-size scale operations.

Tilapia Breeding Made Easy

Breeding tilapias is relatively easy once you have a pair. The natural way to find a breeding pair is to raise several young tilapias together and observe them as they pair up. Feed them well, maintain water quality, provide the right conditions, and they will spawn readily.

Most tilapias are open substrate spawners who dig large pits in which the male and female will clean and prepare before laying their eggs. The female will select her spot to lay her eggs and the male will follow behind to fertilize them.

The female will use her ovipositor, which is a short, wide tube that dispenses her eggs. The male uses his own ovipositor, which is longer and thinner than the female's. He uses this ovipositor to fertilize the eggs with his sperm.

Tilapia parents have the tendency to be extremely aggressive toward all other fish when spawning. If spawning several pairs in the same tank, be sure to add plenty of hiding places.

Large quantities of fry are produced in each brood. Much like many other cichlids, tilapias use their great parenting skills to protect and provide for their young. Tilapia fry grow quickly and can begin eating as soon as their egg sacs are absorbed. In just a few months, these fry will be able to produce their own young.

Tilapia Breeding Methods

A male and female tilapia pair can be separated out to have their own tank, or several pairs can be bred in the same tank, so long as there is an abundance of space for each pair to have their own territorial area. Place a flowerpot (laying on its side) into the tank with the open end of the flowerpot facing the tank wall, about 8-inches away from the tank wall.

The male will make his territory between the open end of the pot and the tank wall. This allows females who are on the other side of the flowerpot to be out of site when he is in his territory. The male tilapia claims a territory usually 2.5 times his body length in all directions that he can see from the center of his arena.

Place some pea size gravel in and around the flowerpot before adding the fish to the tank. Be sure to rinse the gravel before adding it to the tank. Make sure the tank also has the preferred water quality before adding the fish. The temperature of the water needs to be within the desired range and similar to the tank temperature where the fish are being removed from.

Tips from Commercial Tilapia Breeding Professionals

If breeding several pairs of tilapias within the same tank, it is wise to take a page out of the professional breeder's playbook. Professional tilapia breeders cut the upper lip off the male with a sharp pair of scissors before adding them to the breeding tank. Pulling the upper lip out will reveal the articulation between the edge of the lip and the front area of the tilapia.

The reason for trimming his lip is that the male is extremely aggressive with any fish that is within the area he claims as his territory. Breeding multiple pairs of tilapias in a tank often leaves the females with nowhere to get away from the breeding efforts of the males. The constant harassment from the males can be fatal to other fish.

The flowerpot and anything else that you can add to help define his area, such as a small piece of plywood standing vertically near the base of the pot, is helpful, but it typically does not completely prevent the male tilapia from continuously chasing the females, even if they are not ready to breed. The male bites them and after so much of this, he ends up scraping a lot of skin and scales off of them. Removing the male tilapia's upper lip prevents him from causing injury to other fish but does not interfere with his ability to breed with the females that are ready.

To perform this procedure, take the scissors while holding the male firmly (with a towel or wash cloth), then cut a line across the hinge of the upper lip through the thin membrane and the center cartilage over to the opposite hinge, making a clean cut. This cut heals quickly, and the male is capable of breeding within a few minutes of being placed with the females. Once the males are trimmed, they can be placed in the tank with the females.

Breeding Tank Conditions

Automatic controls need to be set up for the breeding tank that maintains the optimal conditions for successful breeding.

1. **Lighting:** If you are in a closed room rather than an open greenhouse, the lights should come on at 6am and go off at midnight. If you place the tank in a greenhouse or near a brightly lit window, then the lights should come on at 6pm and go off at midnight.
2. **Heating:** Plug in an aquarium heater with a thermostat. Set the thermostat 85 to 88 degrees F (29 to 31 degrees Celsius).
3. **Thermometer:** A thermometer needs to be installed so that the temperature of the tank water can be checked at least once a day.
4. **Oxygenation:** Plug in an air pump and attach a line to the filter and air stones, which are located in the tank as faraway as possible from where the flowerpot(s) is placed. Once a day, the air pump should be checked to ensure the bubbles are being delivered out of the top of each filter and that the air stone is properly connected to it. If the oxygen level is low, the Tilapia will usually warn you in advance by rising to the surface and skittering. The filter should be taken out of the tank about every three days and washed out under running

water and then put back into the tank and hooked up to the air.

Feed for Breeders

Professional tilapia breeders claim that they have better success using a type of feed for breeding that is different from that used to grow or maintain tilapia. The feed used by commercial tilapia breeders has a higher concentration of protein and vitamins, which better meets the needs of the breeders for producing healthy eggs and sperm. The feed also needs to have a higher utilization rate to reduce the amount of waste so that cleaning up after the breeders is not a constant chore.

Commercial Tilapia breeders usually use a good quality salmon or trout breeder chow booster. Other commercial breeders use Tuna Tender Vittles cat food from Purina.

Feeding Process for Breeders

Feeding should be done two or three times a day. Start off using a teaspoon of feed for every ten tilapia each feed session. If all the feed is being consumed, then add a little more each feed session until you observe that there is left-over feed. This will be the threshold and the point in which you will know the correct amount of feed to use each feed session.

Breeder Transfer

As the male dances around a female and she joins in the dance, it is a good indication that they will breed soon. Once they have bred, the female will have a mouthful of eggs and will not show much interest in eating. When a female is observed in this state, a note should be made and then in three or four days, she should be gently removed from the breed tank and placed in a separate tank until her eggs have developed into free-swimming fry. The best way to do thisis to take two 6 by 6-inch nets and very slowly herd her into one. Gently hold the net against the side of the tank wall and bring it up over the side, transferring her immediately into a bucket with water that was taken directly from the breeder tank. Quickly and gently, transport the female into the tank exclusively for her. This is referred to as the nursery tank. The nursery tank needs to have the water at the same temperature as the breeding tank.

Leaving a female in the tank after she has bred will result in fewer eggs reaching the hatching stage. Other female adults will eat many free-swimming fry that do develop from the remaining eggs. Therefore, to obtain maximum breeding results, having a separate nursery tank especially important.

If the female spills her eggs during the transfer, don't fret. Although this is not ideal, all is not lost. Catch the female, preferably with two nets, and transfer her and her eggs to the nursing tank. If she has eggs in her mouth and spits them out, simply pick them up with the net and transfer those eggs as well. Once the female and her eggs are moved, she will usually pick them up relatively soon and continue the incubation process.

Sometimes, especially with young females, the male does not properly fertilize the first brood, and the female will not swallow them because she instinctively knows that they are not developing properly. Each female incubates her eggs by slowly and continuously rolling them gently in her mouth. She can tell whether each egg is sick or dead, and then separate those eggs from healthy live eggs while swallowing any that are not right. The eggs begin development almost immediately after the female picks them up in her mouth and within 48 hours, the beginnings of eyes and tails can be seen on the eggs. By the fourth day the fry begin to resemble small fish attached to little yellow balls, which are the egg sacks.

Post-Breeding Behavior

Tilapia males in general are aggressive and territorial. To a certain extent, the female tilapia also becomes territorial for a period after breeding. Besides the obvious full mouth, the female develops a dark marking on her forehead and the vertical stripes on her body become darker when brooding eggs. She also becomes more sensitive to many things and can be upset rather easily.

Nursery Tank Management

When a female is stressed in the nursery tank, she may get overly excited and try to escape. This may cause her to spit all or part of her eggs out. The best thing, of course, is to move very slowly and gently when near the tank. The more you can do to minimize unnecessary disturbances, the better. For instance, don't play music, slam doors, clang pipes and avoid turning lights off and on unnecessarily. This is being a responsible, animal friendly operator.

Tilapia Motherly Behavior

The mother tilapia behavior changes as the young eggs develop into fry (baby fish). During the first two or three days, she simply swims around in a group with the rest of the females and young males. By the fourth or fifth day, she begins to look for a place that she can claim as her home area. During this time, she turns a darker shade on the front of her head and sometimes develops darker bands running vertically (kind of like a zebra). Scientists believe that this darker forehead is a warning signal to other tilapia and fish to stay away, because if they make the mistake of coming too close to her, she will dart at them and even bite them if she can.

By the fifth to ninth day, the fry can navigate well enough on their own to be released by the mother for brief excursions out of her territorial area. At first, she lets them out for a brief swim and sucks them back in within a few minutes. By the sixth day, she allows them to feed on zooplankton, bacteria, algae, and fungal growths on the surface of plants, rocks, or tank walls. While the fry are free swimming, the mother tilapia keeps a sharp eye out for any intruders such as other fish and will aggressively chase them away if they approach the area where the fry are feeding.

If the mother perceives any danger to the fry that she cannot chase away, she will signal the fry by with a sideways wriggle of her body and an open mouth. The fry will immediately swim towards and into her mouth.

The older the fry get, the more time the mother allows them to spend outside. By the tenth day, she will no longer tend them or allow them oral sanctuary. It is also true that the older the fry are, when the mother is disturbed, the more likely she is to spit them out to fend for themselves if she feels her life is in danger.

Tilapia Mother Post Nursing Phases

Once the mother tilapia begins to allow the fry out to browse on microscopic plankton, the mother can be caught and put back into the breeder tank to breed again. However, for ethical reasons, I believe it is more appropriate to wait unto the mother and fry are more comfortable about being on their own, generally two to three weeks after birth. It is best to catch her with a net with at least 1/4-inch holes in it so as to allow any fry to escape. When picking up the mother tilapia, hold her gently and firmly with a wet towel or a cloth glove. It is also advisable to place your fingertip on her lower lip and pull downward to open her mouth to ensure she is not harboring any fry in her mouth.

If she is holding fry, continue holding her mouth open and place her back into the tank or into a large shallow pan with water from the tank and swish her forward and backward until all of the fry are washed out. You may then place her back into the breeding tank and return to the fry.

Tilapia Fry to Fingerlings

Immediately after the mother is removed from the nursing tank, fry frantically search for a hole or nook that will take the place of their mother's mouth. This behavior typically lasts a few hours. To be kind to your animals, and reduce their stress, place something in the nursing tank (i.e., rinsed concrete blocks with holes, thick leafless branches, etc.) a couple of days in advance of taking the mother out so that the fry can take refuge and feel secure.

Feeding can begin on their first day, but keep in mind that they will not consume large quantities of food, or some may not eat at all. Some tilapia fry may not eat the first day or two because the yolk sac egg is still showing in their ownstomachs. Any feed that is not eaten by the second feeding should be taken out of the tank and disposed of. After day four, for best results, feed the fry three times a day. Continue to keep the tank as clean as possible. An aquarium cleaning suction device works great for cleaning the tank.

By the second or third day, the fry will swarm around the feed and gobble it up with astonishing tenacity. Feed them as much as they will eat within a fifteen-minute period. Once you know about how much they eat on an individual feeding, try to give them that much for the next two feedings that day. Each day they will eat a little more as long as the oxygen levels, temperature, pH, and water quality parameters remain within the tilapia's desirable range.

Tilapia Fry Food

The food for the fry needs to be either live food like zooplankton, brine shrimp or high protein powder of flake food such as is used for trout or salmon fry. Obviously, the closer one can stay to a healthy and organic food source, the better.

Many operators claim that the best food they have found is a diet consisting of a mixture of dried spirulina and artificial zooplankton. This type of feed can be purchased online or from most local tropical fish stores. If your local store does not carry it, you can ask them to order it for you. This approach will often enable you to acquire the feed you are after without having to incur the shipping cost.

There are also a number of other ways to feed tilapia fry. One method is to take a quart of the same food that you use for the larger tilapia and soften it in water. Next, add two eggs and blend it until it becomes soupy. Mix it with two cups of boiled water and stir-in one ounce of *Knox Gelatin*. Place the mixture into a bucket, pan, or bowl and refrigerate it until needed. When feeding it to the fry, drop a half-teaspoon into the tank for each feeding.

Moving the Fry out of the Nursing Tank

When the fry reach about 1-inch (2.54 centimeters) in length or more they can be moved into a larger growing area, such as a pond, a larger tank.

Tilapia Production Goal

How many pounds of fish do you want to have each week? The amount of tank space you can provide for growing fish can then be designed to produce the amount of fish that you wish to harvest or eat each week. The production per week goal is usually figured by the cubic foot of growing space—defined by the size of the tank—and the type of system being used. These factors determine how many fish can be stocked.

For instance, in a small tank or a tank with negotiable water exchange, the weight of fish that can be maintained is exceedingly small (around one to two ounces of fish per cubic foot). This type of set up is precarious because the amount of delicate balance of life supporting parameters. A minor disruption of dissolved oxygen levels, a little left-over feed, and/or waste generated can easily upset the water quality conditions. Therefore, fish density must be kept to a minimum to ensure life support conditions are not compromised.

If an aeration device is added, a much higher density of fish can be achieved. This will enable production to be increased substantially, with up to a half pound of fish per cubic of foot water being possible.

For example, considering a tank that is 10 feet by 10 feet with 3 feet of depth is 300 cubic feet. A tank with limited aeration can only hold one ounce per cubic foot of water and would only achieve a production amount of

300 total ounces (18.75 lbs.) of fish in the tank at any one time. Having this same tank integrated into an aquaculture system, producing a half-pound of fish per cubic of foot water, would result in a total tank production of 150 pounds at any one time. That is a significant difference of 131.25 lbs. (59.5 kg). In a well-designed system, the media bed serves as an effective means of eliminating waste and increasing dissolved oxygen.

Tilapia Weight Gain and Measurement

Per industry nomenclature, a standing crop is the total weight of all of the fish in a tank or pond at any chosen moment of time. This number is generally expressed in pounds per cubic foot.

It is important to note that most literature discussing the rearing, breeding, and harvesting of tilapia is based upon large commercial scale operations. These operations typically discuss tilapia in regard to larger scale measurements pertaining to ponds referenced in acre-feet or hectors. Fret not, for the oxygen required per cubic foot, the pH, dissolved oxygen parameters, temperature, feed required per cubic foot, feed conversion rates, as well as other life sustaining parameters referenced will be the same for a small hobbyist size fish tank as it is for a large commercial aquaculture pond.

The growth rate of tilapia is determined by several factors. It is affected by water quality, temperature, oxygen levels, and the general health of your fish. The type of food, along with the quantities provided, is also of imperative importance. Ensuring stocking density does not exceed the optimal range is a critical factor as well.

Furthermore, it is important to choose a species, hybrid, or strain, that is a good 'fit' for your particular operation and goals. Many tilapia vendors advertise strains with a super-fast growth rate. However, the purported growth rate will not be attained unless the living environment is ideal and suitable for their particular needs. The tank environment must be considered.

Mixed-Sex versus Mono-Sex Culture

When male and female tilapias are kept together, they will readily breed and produce a lot of offspring. This will hamper the growth rate of the adult fish, as they will be forced to compete for food with fry and fingerlings. The three following methods are commonly utilized to prevent this from happening:

1. Harvesting the mix-sexed culture before they reach sexual maturity or soon afterwards.
2. Raising the mix-sexed culture in cages or tanks that disrupts preproduction.
3. Raising a mono-sex culture consisting of males only.

Growth Rate in Mixed-Sex Culture

In a mixed-sex tilapia culture, fish are typically harvested before they reach sexual maturity or soon afterwards. This restricted culture period makes it even more important than normal to facilitate the fish growing as quickly as possible since they have to reach their proper size within a limited time frame. It is therefore common to avoid dense stocking of mixed-sex tilapia cultures. It is also important to avoid using stunted fish since such fish will reach sexual maturity while they are still too small for the food market.

Blue Tilapia (*Oreochromis aureus*), Nile Tilapia (*Oreochromis* **niloticus**), and their hybrids are common in mixed sex

cultures since they will attain a marketable size before they commence to spawning. Species such as Mozambique Tilapia (*Oreochromis mossambicus*) and Wami Tilapia (*Oreochromis urolepis hornorum*) are avoided by most operators since they will be too small when they reach sexual maturity.

By choosing the right strain of tilapia and providing the fish with a suitable environment and proper nutrition, it is possible to achieve a growth rate fast enough to allow fry produced in the spring to reach an editable (marketable) size by autumn in temperate regions. For a four to five month long culture period, it is common to stock one month-old fry in grow out tanks. The average weight at harvest can then be expected to be around 0.5 pounds (220 grams), when supplemental feedings with protein rich food is carried out. The recommended stocking density is one pound of fish per five to seven gallons of tank water (.5 kg per 20-26 liters). Therefore, for the most part, the quantity that can be raised is largely dependent upon the tank size.

Tilapia Growth Rate for All-Male Fingerlings

All male fish are grown by operators using a mono-sex cultures method, since the male tilapia grows faster and reaches a larger size than the female. All male batches can be obtained through hybridization, hormonal treatment, or manual sexing and separation. It is important to note that none of these methods can guarantee 100 percent males in every batch. If you desire large tilapia, the number of females in the growing unit should not exceed 4 percent. Many operators use more than one method to ensure a low degree of females in the growing unit. Predator fish of a suitable size can also be added to the growing unit to devour any offspring.

All-male tilapia cultures are often densely stocked. This decreases the individual growth rate of each fish, but it normally results in a higher yield-per-unit area. Densely stocked cultures are more susceptible to ill-health, so careful water management is critical since poor health can have a devastating effect on growth rate and lead to massive losses. In a suitable environment with an adequate supply of nutrition, it is possible for 50-gram fingerlings to become 500-gram fishes within six months. This means an average growth rate of 2.5 grams per day. You can expect the average weight gain to be 1.5-2.0 grams/day. The culture period needs to be at least 200 days, often more, if you want to produce fish that weighs almost 500 grams. Keep in mind, that maintaining water quality at higher densities becomes more challenging.

Table 26 Below: Average production values for male mono-sex Nile and Red Tilapia in system. Nile Tilapia are stocked at 0.29 fish/gallon (77 fish/m3) and Red Tilapia are stocked at 0.58 fish/gallon (154 fish/m3).

TABLE 26. **Average production values for male mono-sex Nile and Red Tilapia in an aquaculture system**

HARVEST WEIGHT PER TANK (lbs)	HARVEST WEIGHT PER UNIT VOLUME ((lb/gal)	INITIAL WEIGHT (g/fish)	FINAL WEIGHT (g/fish)	SURVIVAL (%)
1,056 (480 kg)	0.51 (61.5 kg/m3)	79.2	813.8	98.3
1,212 (551 kg)	0.59 (70.7 kg/m3)	58.8	512.5	89.9

PART IV : OPERATION AND MAINTENANCE

Aquaculture in North Carolina

Tilapia

Inputs, Outputs and Economics

North Carolina Department of Agriculture and Consumer Services

Aquaculture and Natural Resources

Aquaculture in North Carolina – Tilapia

Contents

~~~

About This Publication 2
U.S. Aquaculture and Tilapia Production 2
North Carolina Tilapia Production 2-3
An Example System: The NC State Fish Barn 3-4
Inputs 4
    Water 4
    Land & Buildings 4
    Equipment 4-5
        Oxygenation System 5
        Waste Removal 5
        Feeding 5-6
        Harvesting & Grading 6
        Instrumentation & Control System 6
        Medication & Chemicals 6
    Electricity 7
    Fingerlings 7
    Labor 7
Outputs 7
    Tilapia 7
    Effluent/Compost 7
Economics 7-8
    Initial investment 8
    Operating costs & returns 8
Other Topics 9
    Financing 9
    Farm size 9
    Insurance 9
    Permits & licenses 9-10
    Alternative uses 10
    Markets 10
    Research 10
Tilapia Budgets 11-15

Aquaculture in North Carolina ~ Tilapia

About this publication
~~~
The North Carolina Department of Agriculture, Division of Aquaculture and Natural Resources, created this publication to assist individuals interested in the business of tilapia farming. The publication was also designed for bank lenders who may need more information on the industry to evaluate loan proposals. A description of the inputs and outputs of the North Carolina State University Fish Barn, as well as an estimate of costs, returns, and resource requirements are provided. For technical recommendations on building and operating a fish farm, individuals are encouraged to contact agents with the North Carolina Cooperative Extension Service. For information on state regulations governing aquaculture, or for help in preparing an aquaculture business plan, contact the North Carolina Department of Agriculture (see Sources of more Information on page 16).

### U.S. Aquaculture and Tilapia Production
~~~

Aquaculture is the fastest growing segment of U.S. agriculture. The farm-gate value of the U.S. aquaculture industry is estimated at nearly $1 billion. Tilapia food fish production accounts for about 4% of the total value.

Tilapia are a group of two dozen species of freshwater, plant-eating fish native to Africa. Tilapia have been cultured for thousands of years, are the second largest aquacultured species in the world (after carp) and are grown in over 80 countries. They are considered an ideal fish for aquaculture; tilapia grow in a variety of environmental conditions, breed easily in captivity, are efficient converters of feed to fish meat, and have a mild flavor which appeals to consumers.

Production of tilapia has been one of the fastest growing aquacultured species in the U.S. In 2001 an estimated 20 million pounds were grown by over 100 producers. Most tilapia is produced in one of six large US facilities, each of which grows over one million pounds annually. In addition, there are about 20 operations producing from 200,000 to one million pounds, and 50 to 100 producing less than 200,000 pounds annually.

Most US producers grow the fast-growing species *Oreochromis niloticus*. While the vast majority of tilapia grown in the world are raised in ponds in tropical climates, pond culture in the U.S. is only feasible in the southernmost parts of Texas and Florida, or where geothermally heated water is available. Tilapia thrive in waters having a temperature of 82..to 86..F. Tilapia grow much more slowly at a lower temperature and lose their resistance to disease when temperatures fall below 70..F. Tilapia cannot survive in water below 55…F, precluding their survival outdoors in North Carolina.

Ninety percent of U.S.-grown tilapia are raised in water-reuse or so-called recirculating systems. These systems raise fish in tanks, usually indoors. Specialized equipment is used to filter waste from the water and add oxygen as needed. Nearly all of the water is reused rather than discarded.

North Carolina Tilapia Production
~~~

North Carolina has grown rapidly in the past few years as a source of U.S.-grown tilapia. The state has one integrated

Aquaculture in North Carolina ~ Tilapia

hatchery, grow-out, and processing operation with a capacity to produce over 500,000 pounds of tilapia per year, and 10 operations with a stated design capacity of 200,000 to 400,000 pounds each in existence or under construction. There is also a commercial-size research and demonstration facility, the NC State Fish Barn, located at North Carolina State University in Raleigh. Tours and classes are held in conjunction with the facility. All North Carolina tilapia production facilities are indoor water-reuse systems.

An Example System: The NC State Fish Barn
~~~

To make tilapia production economically feasible in an enclosed (indoor tank) system, fish are grown at much higher densities than they would grow in a natural setting. For example, pond-raised catfish are raised at a rate of 0.005 to 0.007 pounds of fish per gallon of pond water. In contrast, the example system presented in this publication allows a density 100 times as great: tilapia are grown at a maximum density of 0.66 pounds of tilapia per gallon of water. The sophisticated water treatment components in tank systems allow for this density of production.

There are dozens of different types of tilapia tank systems in use around the world. While there is no single recommended or standard tank system, there are some basic components that the system must have. Each system is composed of a combination of tanks, pumps and filters of varying size, shape, and efficacy. The costs of building and operating each system and the returns to each differ markedly. The costs and returns given in Tilapia Budgets (pages 11-15) and explained below are valid only for facilities based upon the model at the NC State University Fish Barn. Potential producers are advised to contact researchers at the Fish Barn (see Sources of More Information) for assistance in setting up their own tilapia production system or in evaluating a specific design.

The example Fish Barn is a 32 x 130 insulated building with six fiberglass tanks, varying in size from 1,800 to 15,000 gallons, plus associated monitoring, harvesting, and water treatment equipment. Up to 10% of the water volume of the system is exchanged each day (that is, 10% new water enters, while 90% is filtered and reused). The discharged water goes through a treatment system to a 1/2-acre effluent pond.

Incoming fingerlings are initially stocked into an 1,800 gallon quarantine (Q1) tank, where they are held and screened for diseases for 6 weeks. They are then harvested and restocked into the 4,000 gallon nursery (Q2) tank. After another 6 weeks of growth the fish are transferred to one of the four 15,000 gallon grow-out tanks (G tanks). After another 6 weeks, half of the fish are stocked into another of the G tanks. There the fish remain for another 50-60 days, and then they are harvested. Thus, the total cycle time between first stocking and first harvest is about 180 days. One tank is harvested per month, in four equal weekly batches, with 2,389 pounds harvested each week.

Fish are moved from tank to tank to make optimum use of the production capacity of the system. As the tilapia grow and require a larger water volume, they are transferred into larger tanks. When the preceding tank is unoccupied, another group of fish called a cohort is introduced into the system. As a result, once the system is fully stocked, one G

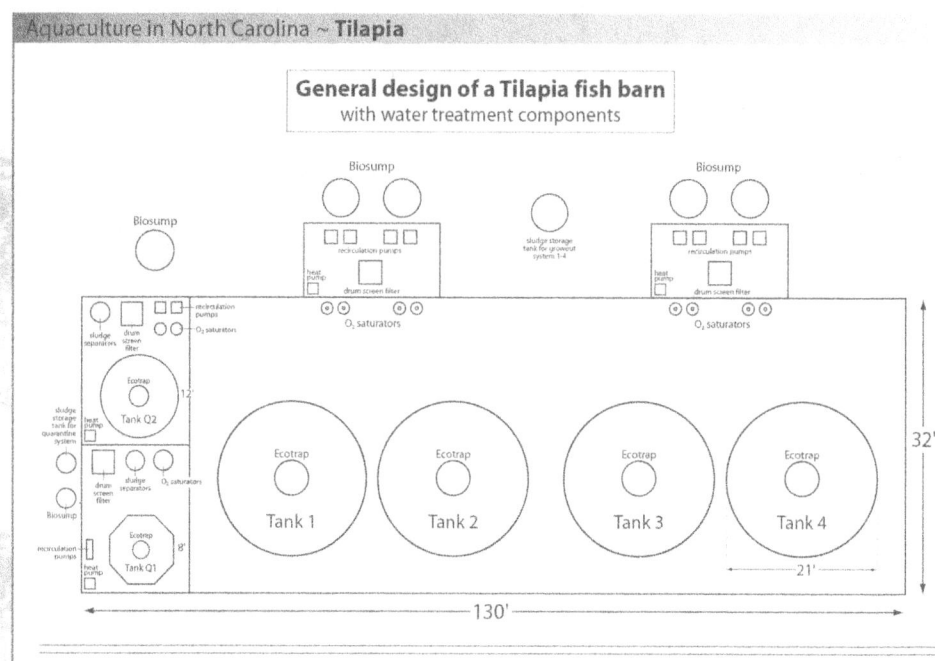

tank will be harvested every 180 to 210 days, resulting in a constant, year-round supply of tilapia.

Inputs
~~~
#### Water

A major advantage of water-reuse systems is that water is not usually a limiting factor when siting the facility. The example facility has a well with a pumping capacity of 35 gallons per minute (gpm), but generally runs at a constant rate of just 5 gpm. Water must be derived from a well (as opposed to a surface water source such as a stream) to bar introduction of sediments, other organisms, and pathogens. The facility exchanges about 10% of the total tank volume per day, requiring 6,600 gallons per day. At each harvest, tanks are partially drained, but nearly all of the new water volume is saved in the biosumps (biofilter) and another tank. A flow meter for each system measures the amount of new water added to each separate water reuse system.

#### Land and buildings

Land is also not usually a limiting factor for siting a water reuse system, as so little is needed and virtually any soil type may be used. Tilapia Budgets assume a 5-acre tract used for the building and 1/2-acre effluent pond which holds water discharged from the system. A ß hp aerator runs nightly in the late spring, summer and early fall to insure that the pond contains enough oxygen.

The building is a 32 by 130 insulated metal-clad pole barn. Tilapia Budgets separate the costs of the actual building from plumbing and electrical systems and the labor to install these systems, because owner/operators often supply their own labor.

#### Equipment

There are four separate recirculating systems: one for the Q1 tank, one for the Q2 tank, and one for each of the two pairs of grow-out tanks. Systems are separated to prevent pathogen transfer between the systems. It is particularly important that

Aquaculture in North Carolina ~ **Tilapia**

### *Water Treatment System Components*
~~~

Recirculation pumps create a constant flow of water from the tanks, through the filtering and oxygenation systems, and back to the tanks

Sludge collector	waste solids removal
Particle Trap (Ecotrap)	waste solids removal
Drum screen filter	filters out waste solids from the tanks (fecal matter, food particles) for further treatment
Biosump (biofilter)	ammonia control, carbon dioxide stripping, re-aeration of system water
Oxygen saturator	transfers oxygen into the water
Heater (electrical and gas heaters)	maintains water temperature at a constant 84…

disease problems be detected and treated early in the cycle, and that they not spread to the entire facility. For this reason the Q1 and Q2 tanks are isolated in their own rooms with separate recirculating systems, even though this increases the costs of the facility. Grow-out tanks are paired because the probability of disease is less later in the grow-out cycle than it is at the quarantine and nursery stages.

Each of the four systems has its own recirculation pumps, drum screen filter, biosump (biofilter), particle trap (Ecotrap), sludge collector, and oxygen saturator. The table above gives a brief description of the function of each component, and the costs per component are given in Tilapia Budgets (pages 11-15).

For each system, the water flow is from the tank through the particle trap s main discharge line. The sludge output from the Ecotrap is routed to the sludge collector, where the sludge particles are removed before passing the clarified water through the drum screen filter to the biosump. Return from the biosump is through the main pump suction line, main pump, oxygen saturator, and water heater, and back into the tank.

Oxygen System

The oxygenation system is comprised of a bulk liquid oxygen storage tank, oxygen saturators and in-tank emergency oxygen supply diffusers. Oxygen is continuously supplied to the tank, even during a power outage. The liquid oxygen storage tank is provided by a local supplier and is re-filled from a tanker truck when the oxygen supply reaches a predetermined level.

Waste removal

The waste removal system consists of in-ground sludge storage tanks, sludge pumps, a compost facility for disposal and treatment of solid materials, a settling tank, and a pond for discharged water.

Feeding

Automatic feeders distribute feed each ß hour for fish less than one pound, and each hour for fish greater than one pound. The uniform, timed application of feed throughout the day results in a more stable daily water quality pattern, requiring little

Aquaculture in North Carolina ~ Tilapia

or no manager maintenance control of the oxygen supply. The operator monitors the sludge collectors for excess feed content and, combined with visual observation of feeding, adjusts feed rate and amount in the automatic feeder.

Floating feed is preferred as it allows the fish time to eat the feed, then sinks to the bottom where it can be removed automatically from the system. Feed varies in size and protein content, and each cohort is fed four different types of feed during its growth cycle. Feed is the largest single operating cost, at 20% of the total.

Harvesting and Grading

The equipment used in harvesting and grading the fish consists of nets, baskets, an electric hoist with electronic scales, a custom-built crowder/grader, and an overhead rail for the hoist which runs along the ceiling and down the length of the grow-out tanks.

Instrumentation and Control System (ICS)

The ICS allows for automation of some operations at the Fish Barn, thus reducing the need for labor. The ICS is comprised of various sensors, controls, and alarms. The system utilizes programmable logic controllers to monitor specified parameters (water flow, temperature, oxygen level alarm status, power status, water level) and initialize action (automatic or alarm) based on the specified setpoints. Personnel are called by telephone after hours if there is an emergency, and they can dial in and check the alarm status of the system. In Tilapia Budgets, the cost of the control and alarm system is included in the building electrical costs and telephone dialer.

Medication and Chemicals

Aquaculturists maintain that the key to successful fish management is stress management. Fish can be stressed by changes in temperature and water quality, by handling, and by nutritional deficiencies. Stress increases the susceptibility of fish to disease, which can lead to catastrophic fish losses if not detected and treated quickly.

Disease-causing organisms are likely to enter the system from hauling water, on fingerling fish introduced into the system, or on nets, baskets, gloves, etc., that are moved from tank to tank. Tank facilities guard against the introduction of disease from the outside by screening incoming fish in a quarantine tank, proper cleaning and disinfection of baskets, gloves, etc. used during transfer and harvest of fish, and requiring foot baths and other biosecurity measures for workers and visitors entering the facility. These measures, and the maintenance of a healthy growing environment through proper feed, temperature, and water quality management can keep a Fish Barn virtually disease free.

Three chemicals are added to maintain optimum water quality in the facility.

Rock salt is added to the water on a weekly basis or as needed to maintain chloride levels at 200-300 ppm. Calcium carbonate is added as needed to maintain calcium hardness at a level of 80-100 ppm. Sodium bicarbonate is added twice daily to maintain alkalinity and buffer against pH variations. The total cost of chemicals comprises less than 5% of total variable operating costs.

Additional salt, ozone, or the application of ultraviolet light may be added to the water to reduce a high bacterial count (which may result from over-feeding).

Unchecked, a bacterial or disease problem can quickly result in the loss of an entire tank. It is of the utmost importance that the operator take the measures necessary to keep the facility disease-free.

Electricity

Water reuse systems require significant amounts of energy to operate. For the example system, about 15% of the total electricity is required for the building heating, ventilation, and air conditioning (HVAC) and the remainder for operation of pumps, filtration, and water heating equipment. The total cost of electricity comprises approximately 12% of variable operating cost.

Fingerlings

The quality and health of the incoming stock of fingerlings is of utmost importance. One of the most important factors in procuring quality fish stock is to buy from reputable, established hatcheries. If disease organisms are accidentally introduced into the system, they can be extremely difficult to eradicate or control, requiring the breakdown of the system components for disinfecting and drying. This disrupts the production cycle and can easily result in a net economic loss for the system.

Tilapia fingerlings are readily available from one hatchery in North Carolina and a number of hatcheries across the U.S. The quality of the tilapia, both in terms of genetics and health of the fish, can vary widely between source hatcheries. Operators are advised to deal with reputable hatcheries and refuse to accept questionable loads. Tilapia Budgets includes stocking of 15,000 fish per cohort, one cohort (one tank) stocked each month, for a total of 180,000 fish per year. The cost of fingerlings is about 13% of total variable cost.

Labor

Daily maintenance of the system requires water quality testing as needed, filling of the feeders, addition of sodium bicarbonate and rock salt, cleaning of the piping and equipment (such as the drum screens), visual verification that all systems are acting normally, and recordkeeping. Stocking, sampling, and harvest of the tanks is also required periodically. This requires a relatively high degree of competence. The person in charge must be able to recognize problems in the system that might not be obvious to an inexperienced person.

The example Fish Barn is designed to minimize operator labor. Eight hours per day are typically required, with less time on weekends. About 2,400 pounds of fish are harvested one day each week, and this requires another part-time laborer. Fish can also be transferred on this day. Tilapia Budgets do not include the cost of the owner's labor. They do include $7,680 annually for transfer and harvest labor.

Outputs
~~~
### *Tilapia*

The example system has an overall survival of 91%, for an annual harvest of 114,660 pounds of 1.40 pound tilapia. Tilapia Budgets assume that 100% of the fish are sold to a live market. (See Marketing, below).

### *Effluent*

The tilapia operation discharges about 6,600 gallons of water per day, to the effluent pond. Solid wastes flow automatically to storage tanks, which are emptied and pumped to the compost

Aquaculture in North Carolina ~ **Tilapia**

facility weekly. The facility generates about 800 cubic feet per year of compost, which could serve as an additional source of income.

### Economics
~~~

Tilapia Budgets estimate initial investment, operating costs, and annual returns for a 6-tank facility. Production, costs and sale price are based on the experiences over the past 3 years at the North Carolina Fish Barn Project at NC State University.

Initial Investment

The farm requires an initial investment of $301,575. In this example the owner supplies 25% of the total investment, and the remainder is borrowed over a 10 year loan term.

Summary of Initial Investment	
Land	$8,000
Waste removal	$15,000
Building	$61,858
Plumbing and electrical equipment for the building	$19,795
Labor for plumbing, electrical, and equipment setup	$40,000
Equipment	$152,922
Well	$4,000
Total	**$301,575**

Operating Costs and Returns

Sales are based on a price of $1.40 per pound of fish sold. Sale prices for tilapia vary widely by year, season, and source of product, and are in a constant state of flux and uncertainty (discussed below, in Marketing). The $1.40 is based on the experience at the NC Fish Barn, and is based on sales of live fish.

Costs are split into the categories of variable costs and fixed costs. Variable costs vary directly with the volume of output; if nothing is produced, variable costs are zero. Variable costs include the inputs described above (fingerlings, feed, electricity & fuel, etc.), repair and maintenance, and an interest cost on operating capital. Tilapia Budgets assume that the farm finances variable costs with credit lines from the feed mill and bank at an annual interest rate of 10%.

The facility harvests 6 tanks in Year 1, for a total of 57,330 pounds. Because the business has 6 harvests, but must pay for the production cost of 12 tanks, the facility loses $45,130 in year 1. In year 2, there are 12 harvests, for a total of 114,660 pounds. Assuming a sale price of $1.40, the farm earns $25,919. This is the return to the cost of the owner s labor and use of the owner s $70,000 in initial capital.

For the facility to be more profitable, tilapia must be sold at a higher price or they must be raised for a lower cost. Some tilapia producers are able to sell live tilapia at prices of higher than $1.40 per pound to niche markets. In fact, these niche markets are what sustain most of the tilapia facilities in the U.S.. For planning purposes, however, a price higher than $1.40 per pound should not be used unless the proposed facility has identified and established a market at a higher price.

The second means to profitability is to lower the price of production. One way is to increase the number of pounds produced in the facility, or to use larger components (tanks, filters, etc.) to gain an economy of scale. This could lower the fixed cost per pound of production, thus reducing the overall cost, and increasing profits. For this example, if the farm increased production by 15% (18,000 lbs annually), cost per pound would fall to $1.05, and the owner would earn $45,928 annually.

Other Topics
~~~
### Financing

Tilapia aquaculture in an enclosed system is attractive to both potential farmers and investors (including bank lenders), because, unlike pond aquaculture, the fish can be seen and their growing conditions seem to be much more under the control of the operator than culture out-of-doors. The disadvantage for existing row crop or livestock farmers is that they can not use to their advantage large tracts of land and farm equipment that they already own. For example, catfish aquaculture requires large tracts of flat farm land, as well as tractors and other farm equipment. This can give an existing farmer a significant edge in getting into pond aquaculture. Existing buildings on a farm (such as those used for livestock) can not be readily converted into a Fish Barn; in virtually all cases, it is less costly to build from scratch.

While the climate for aquaculture loans has improved as the industry has grown over the past 10 years, lenders still consider fish farming to be riskier than other farm ventures. A fish barn is a specialized and highly capital-intensive facility that is not easily converted to other uses. While fish farming is considered a form of agriculture, it differs in that fish are sold through seafood marketing channels, with which agricultural lenders are usually unfamiliar. Lenders are unsure where tilapia will be sold, and wonder how the bank would be able to sell the fish (and manage the farm) if the fish farm were to fail.

Until just recently, the aquaculture industry did not have any integrators that is, there were not any companies which contracted with producers as is done in the production of hogs, turkeys, and chickens. Over the past several years, one integrator company has emerged, and is contracting for growers in the southeast. The company provides financing and some inputs and purchases the tilapia when they reach market size.

### Farm Size & Productive Capacity

As noted above in the discussion of operating costs and returns, larger farms with greater production can lower the cost per pound, resulting in a higher net profit. For example, if the tank size (and size of pumps, filters, and other water treatment components) were doubled for the example system, the cost of these components would be more, but they would not be double the cost of the smaller sized components. Larger farms would not generally result in a savings in variable costs, however, as feed, fingerlings, electricity, etc. increase in proportion to the number of fish being produced. There could be some savings in management labor, however.

### Insurance

Private insurance is more readily available for tank farming than for production of fish in ponds. Farmers can obtain insurance for equipment failure, lost production, and losses sustained due to interruption of production.

### Permits and Licenses

The North Carolina Department of Agriculture and Consumer Services grants an aquaculture license for a period of five years. The license is free. Because tilapia is not a native species, written permission

Aquaculture in North Carolina ~ **Tilapia**

from the North Carolina Wildlife Resources Commission is also required of new growers. A capacity use permit is required for well-water withdrawals in some areas. If the proposed aquaculture operation is not in a wetland and does not meet the criteria of both (1) discharging water more than 30 days per year and (2) producing more than 100,000 pounds per year, it is likely that no other permits will be needed. Potential producers are encouraged to contact the NC Department of Agriculture (see Sources of More Information) to learn about situations when other permits may be required.

### *Alternative Uses*

The Fish Barn system has been used to grow hybrid striped bass and yellow perch in addition to tilapia. With modifications, the system could also be used to raise flounder, ornamentals, and other fish. The species raised would have to provide a sufficiently high price and/or low operating costs to sustain the relatively high capital costs associated with the building and equipment.

### *Marketing*

The American Tilapia Association is the main provider of information on tilapia markets in the U.S. Its August 2001 report notes that U.S. domestic production increased about 2 million pounds during 2000 to a total of 20 million pounds. Annual sales were estimated at $34 million. An additional 152 million pounds were imported (reported on a live-weight basis, though the fish enters the US market as frozen whole fish or fresh or frozen fillets). Since 1993, domestic production has increased 60 percent, while imports have increased 500 percent.

The marketing of tilapia is extremely competitive. In warmer climates (Costa Rica, southern China and Taiwan, South America), tilapia can be grown cheaply in outdoor ponds, and imported cheaply to the U.S. This has placed significant and growing price pressure on US producers. While the bulk of imports has been frozen whole fish sold at prices of less than $0.60 per pound, in recent years the largest percentage increase has been in fresh and frozen fillets.

Because of the price pressure on whole fish and fillets, most U.S. producers sell their fish live to ethnic markets. Fish are sold in large shipments to seafood wholesalers, as well as to local restaurants and seafood markets. Producers earn a 20-40% premium on live compared to iced fish. New producers are encouraged to develop local niche markets to obtain the highest price with the lowest transport cost.

### *Research*

Researchers in North Carolina, other states, and other countries are performing research on both tilapia culture and recirculating systems in general. Improving water quality, reducing energy use, more efficiently dealing with waste management, and producing faster-growing strains of tilapia with a higher fillet yield are all high-priority topics. Because tilapia can be produced cheaply in ponds in a number of countries, growers in the U.S. must find methods to grow tilapia more efficiently and at a lower cost in recirculating systems. Of equal importance is creating value-added tilapia products that will appeal to consumers.

~~~

Aquaculture in North Carolina ~ **Tilapia**

Tilapia Budgets
~~~

NOTE: These worksheets provide only general costs and returns estimates to fish farming.
Investment costs in particular can vary greatly and are extremely site specific. Prospective fish farmers should use these worksheets as a guide to obtaining costs specific to their site.

number of tanks	6.0
total water volume	65,500.0
building size, sq feet	4,160.0
fish stocked per cohort	15,000.0
cohorts stocked/harvested per yr in full production	12.0
survival	91.00%
fish harvested per cohort	13,650.0
average size at harvest	1.40
FCR	1.4
avg. length of production cycle in days	195.0
pounds harvested per tank	9,555.0
lbs harvested, year 1 (6 tanks)	57,330.0
lbs harvested, year 2 (12 tanks)	114,660.0
fingerling cost	0.10
kwh per pound of production	2.54
bank credit line interest rate for yearly op. expenses	6%
percent of construction financed by owner	0%
percent of equipment contributed by owner	37%
bank interest rate for construction (10 year loan)	6%
bank interest rate for equipment (5 year loan)	6%
sale price per lb	$1.40

For this set of worksheets:

1. No cost is assumed for owner s labor or for the interest cost of using the owner s personal funds. Labor is estimated at 40-50 hours per week. Harvest and transfer labor is estimated at 64 hours per cohort.

2. Budgets assume that all construction and equipment purchses take place at the beginning of year 1. Loan payments begin in year 1, but sales sufficient to cover cost do not take place until year 2. If the owner does not have another source of income to make payments until sales begin, then additional interest costs will be incurred.

3. The owner funds 25% of the total initial investment, or $70,094, in labor & equipment.

Aquaculture in North Carolina ~ **Tilapia**

TILAPIA BUDGETS
INVESTMENT COSTS
New Construction & Equipment

	UNIT	PRICE ($/UNIT)	# OF UNITS	TOTAL($)
Land	acre	4,000.00	2	8,000
Waste Removal				
settling pond	acre	10,000.00	0.5	5,000
aerator (1/2 hp)	unit	2,500.00	1	2,500
composter	unit	7500.00	1	7,500
Subtotal				15,000
Building				
building, 32'x130' pole barn	sq ft	14.64	4224	61,858
electrical	unit	7,395.00	1	7,395
plumbing	unit	7,200.00	1	7,200
HVAC (heating & cooling)	unit	5,200.00	1	5,200
Subtotal				81,653
Well & 3/4 hp pump (35 gpm)	unit	4,000.00	1	4,000
Per-tank system equipment (detailed equip. list on next page)				
grow-out systems	unit	89,660.00	1	89,660
quarantine 1	unit	11,897.00	1	11,897
quarantine 2	unit	18,771.00	1	18,771
Subtotal				120,328
System-wide equipment				
feed bins	unit	3,000.00	2	6,000
feeders	unit	300.00	6	1,800
feeder controller	unit	510.00	1	510
gas generators	unit	4,200.00	1	4,200
oxygen monitor	unit	5,234.00	1	5,234
hoist, trolley & track	unit	2,000.00	1	2,000
crowder (for harvest)	unit	2,500.00		2,500
misc. harvest equipment (nets, baskets, etc)	unit	1,000.00	1	1,000
2-ton water heat pumps	unit	2,000.00	2	4,000
telephone dialer	unit	350.00	1	350
lab equipment	unit	4,000.00	1	4,000
misc. equipment		1,000		1,000
Subtotal				32,594
Labor				
For building electrical and plumbing				20,000
For equipment set-up				20,000
Subtotal				40,000
TOTAL				301,575

Aquaculture in North Carolina ~ Tilapia

	PRICE ($)/UNIT	# OF UNITS	TOTAL($)
Detail of equipment by tank system:			
For grow-out tanks (two pairs)			
tanks (15,000 gallon fiberglass)	5,000.00	4	20,000
pumps (2 hp)	565.00	8	4,520
particle trap (Ecotrap)	3,674.00	4	14,696
oxygen saturator (90-130 gpm)	786.00	8	6,288
foam fractionator (30 gpm min.)	675.00	4	2,700
bio sump	2,000.00	4	8,000
bio sump media	1,280.00	6	7,680
media blower	213.00	4	852
regenerative blower	562.00	2	1,124
biosump level control	200.00	4	800
drum screen filter	11,000.00	2	22,000
drum filter rinse pump	500.00	2	1,000
Subtotal			89,660
For the Q1 tank			
tanks (1,500 gallon fiberglass)	1,000.00	1	1,000
pumps (1 hp)	565.00	1	565
particle trap (Ecotrap)	1,627.00	1	1,627
oxygen saturator (35-85 gpm)	332.00	1	332
foam fractionator (30 gpm min.)	675.00	1	675
bio sump	400.00	1	400
bio sump media	39.00	6	234
media blower	213.00	1	213
regenerative blower	351.00	1	351
biosump level control	200.00	1	200
drum screen filter	5,800.00	1	5,800
drum filter rinse pump	500.00	1	500
Subtotal			11,897
For the Q2 tank			
tanks (4,000 gallon fiberglass)	4,000.00	1	4,000
pumps (1 hp)	565.00	2	1,130
particle trap (Ecotrap)	2,729.00	1	2,729
oxygen saturator (35-85 gpm)	759.00	2	1,518
foam fractionator (30 gpm min.)	675.00	1	675
bio sump	700.00	1	700
bio sump media	149.00	6	894
media blower	213.00	1	213
regenerative blower	412.00	1	412
biosump level control	200.00	1	200
drum screen filter	5,800.00	1	5,800
drum filter rinse pump	500.00	1	500
Subtotal			18,771

TILAPIA BUDGETS
OPERATING COSTS AND RETURNS
Year 1

	UNIT	PRICE/UNIT($)	# UNIT	TOTAL($)	% OF TOTAL	$ PER LB
Gross Receipts						
tilapia	0	1.40	57,330	80,262		
Variable Costs						
fingerlings	per	0.10	180,000	18,000	14.42%	$0.31
feed	lb	0.18	120,393	21,175	16.97%	$0.37
bicarbonate	lbs	0.16	21,069	3,371	2.70%	$0.06
rock salt	mo	50.00	12	600	0.48%	$0.01
chloride	mo	50.00	12	600	0.48%	$0.01
electrical usage						
pumps and filters	mo	1150.00	12	13,800	11.06%	$0.24
building heat and AC	mo	200.00	12	2,400	1.92%	$0.04
propane	mo	300.00	4	1,200	0.96%	$0.02
oxygen	100 cu ft	0.30	8,719	2,616	2.10%	$0.05
repair & maint. of equip.	mo	300.00	12	3,600	2.88%	$0.06
labor, transfer & harvest	$/cohort	640.00	12	7,680	6.15%	$0.13
office overhead	mo	100.00	12	1,200	0.96%	$0.02
interest on above operating funds	dol.			1,649	1.32%	$0.03
marketing cost	dol.			1,000	0.80%	$0.02
SUBTOTAL, VARIABLE COSTS				78,891	63.21%	$1.38
Fixed Costs*						
payment on land and const. debt	dol.			18,838	15.10%	$0.33
payment on equipment debt	dol.			19,663	15.28%	$0.33
property taxes and insurance	dol.			5,000	4.01%	$0.09
oxygen tank rental	mo	250.00	12	3,000	2.40%	$0.05
SUBTOTAL, FIXED COSTS				46,501	36.79%	$0.81
TOTAL COSTS				125,392		$2.19

*Excludes annual depreciation, estimated at $21,246

RETURNS SUMMARY
Returns to owner's management, labor, and capital

	PER LB	TOTAL
Returns above variable costs	$0.02	$1,371
Returns above total costs	($0.79)	($45,130)
Breakeven price/lb above variable costs	$1.38	
Breakeven price/lb above all costs	$2.19	

Aquaculture in North Carolina ~ **Tilapia**

TILAPIA BUDGETS
OPERATING COSTS AND RETURNS
Year 2 and Thereafter

	UNIT	PRICE/ UNIT($)	# UNIT	TOTAL($)	% OF TOTAL	$ PER LB
Gross Receipts						
tilapia	lb	1.40	114,660	160,524		
Variable Costs						
fingerlings	per	0.10	180,000	18,000	13.43%	0.16
feed	lb	0.18	160,524	28,233	21.07%	0.25
bicarbonate	lbs	0.16	28,092	4,495	3.35%	0.04
rock salt	mo	50.00	12	600	0.45%	0.01
chloride	mo	50.00	12	600	0.45%	0.01
electrical usage						-
pumps and filters	mo	1150.00	12	13,800	10.30%	0.12
building heat and AC	mo	200.00	12	2,400	1.79%	0.02
propane	mo	300.00	4	1,200	0.90%	0.01
oxygen	100 cu ft	0.30	11,625	3,488	2.60%	0.03
repair & maint. of equip.	mo	300.00	12	3,600	2.69%	0.03
labor, transfer & harvest	$/cohort	640.00	12	7,680	5.73%	0.07
office overhead	mo	100.00	12	1,200	0.90%	0.01
interest on above operating funds	dol.			1,808	1.35%	0.02
marketing cost	dol.			1,000	0.75%	0.01
SUBTOTAL, VARIABLE COSTS				88,103	65.74%	0.77
Fixed Costs*						
payment on land and const. debt	dol.			18,838	14.06%	0.16
payment on equipment debt	dol.			19,663	14.23%	0.17
property taxes and insurance	dol.			5,000	3.73%	0.04
oxygen tank rental	mo	250.00	12	3,000	2.24%	0.03
SUBTOTAL, FIXED COSTS				46,501	34.26%	0.40
TOTAL COSTS				134,605		1.17

*Excludes annual depreciation, estimated at $21,246

RETURNS SUMMARY
Returns to owner's management, labor, and capital

	LB	FARM
Returns above variable costs	$0.63	$72,421
Returns above total costs	$0.23	$25,919
Breakeven price/lb above variable costs	$0.77	
Breakeven price/lb above all costs	$1.17	

PART IV : OPERATION AND MAINTENANCE

PART V

Tilapia Business Success

CHAPTER 17

Bartering Your Aquaculture Products

Bartering

Bartering is the process of obtaining goods or services by direct exchange without the use of currency. Bartering is an excellent way to ensure the flow of necessary items and services into your household without using precious funds. Bartering is especially effective in times of economic instability or currency devaluation.

Historically, bartering was conducted through face-to-face exchanges. This method is still common in developing countries and is still conducted, to some extent, in developed countries. The internet has opened up a new medium for bartering opportunities for both person-to-person exchanges and third-party facilitated transactions.

Why Barter?

There are many reasons to participate in a bartering process. Numerous people have found themselves unemployed or with a limited cash flow, and bartering is a great way to attain products and services when times are tough.

Bartering can be done to cut costs of a small business or to reduce personal expenses. For example, aquaponic products can be bartered for carpet cleaning, hairstylist, mechanical work, or other products.

Bartering works especially well between farmers. For instance, an aquaculture operator can swap fish for another farmer's fruit or vegetables. Each person is still obtaining something of value, and it opens up another means for which you can be compensated for your product.

What Can Be Bartered?

Any goods or services that is desired by another per- son can be bartered. Bartering is limited only by one's imagination. For instance, 17-year-old Steven Ortiz made national headlines by bartering a cell phone to start a series of trades that ultimately put him in the driver's seat of a Porsche. Kyle MacDonald bartered his way from a single red paperclip to a house in a series of fourteen online trades over the course of a year. Even healthcare isn't out of reach. Matthew Wagner of Connecticut was able to exchange his photography services for Lasik eye surgery through a barter exchange.

Although these stories, and many like them, are amusing as well as inspiring, one does not have to "trade-up" to be successful. A successful barter can be a trade for anything, in less or more value, that satisfies you. For instance, trading a several of your fish for a sack of fruit can easily be a win-win for both parties. Including cash with a barter for goods or services is also an option. The following are some of the most popular items that can be bartered:

- **Services** — Haircuts, massages, mechanical work, plumbing repairs, landscaping, and a variety of personal care services can be acquired through a trade. Utilizing Facebook Market Place and Craigslist can lead to regular trade arrangements and help build bartering relationships and skills, which can lead to future cash sales.
- **Technology** — Electronic products or repairs.
- **Clothing** — Clothing and accessories. We all need clothes and just about everybody has clothing items that we can do without.
- **Toys and Hobbies**
- **Gifts & Crafts**
- **Food** — Food which you are not growing.
- **Materials & Supplies** — This could encompass anything from building materials, cleaning supplies, toiletries, auto parts, etc.

Bartering Methods

Face-to-Face

Occasionally, you may find bartering opportunities arise while you are with a group of friends, co-workers, or acquaintances. Other times it may be more intentional, such as approaching someone that has something you desire in order to inquire (persuade) about trading for what you have available.

Online via the Internet

The internet has facilitated a unique system of bartering with strangers. There is no need for introductions. In most cases it is as simple as posting an ad describing what you have to offer, and what you seek in exchange.

It is important to always exercise caution when utilizing the internet for bartering purposes. Not every bartering or swapping website, nor every person using these sites, are reputable. Several cities in the U.S. have created a "safe spot" for the exchange of items between people. These areas are monitored by cameras and/or police officers.

Craigslist

Craigslist is well-known website that has a section dedicated for bartering purposes. To utilize this feature, you simply go through the same process you would to post any other item for sale. It is completely free. However, there is no one monitoring the barter ads, so you must be aware of potential Craigslist scams, and realize that you are always at risk when it comes to meetups and exchanges. It is best to meet in a public place, and/or to have someone with you.

U-Exchange.com

U-Exchange is a bartering website that allows people to trade goods and services in a specific geographical area.

The site requires that you register, but it is a free service that is supplemented by banner advertisements.

Just like Craigslist, U-Exchange is a general posting site, and you assume all risk and responsibility for contacts and exchanges you make. When you see a trade that you are interested in making, click on the member's name and you will be provided with their contact information. From there it typically proceeds to a face-to-face exchange.

Cautions of Bartering

There are reasons why bartering is not the dominant system anymore, as it does have some disadvantages, such as the availability of the product or service you are seeking. Simply put, the bartering market is not nearly as liquid as other markets. You may find yourself with something to trade, but no one to trade with at your desired time. This leaves you waiting to make your trade.

Bartering Rules

There are a number of rules for bartering, for reasons of safety and courtesy:

- **Remember, "Safety First."** Meet in a public place or have support with you.
- **Be Inquisitive.** Explore trade options. Remember, it never hurts to ask; the worst they can say is "no."
- **Consider All the Goods and Services at Your Disposal.** A great opportunity can be found by keeping your options open as to what and how much you will trade, and what you will accept.
- **Be Skeptical when Necessary.** Beware of services that do not appear to be legitimate. Also, if you wouldn't pay in monetary value for a good or service, then don't barter either.
- **Don't Barter for Something You Don't Want, Need, or Can't Profit from Later.** You should never trade for something you will later regret. It just simply isn't prudent to trade some- thing of yours that you consider valuable for another's goods or services you deem unnecessary or unwanted.
- **Test Items to Be Sure They Work.** Remember, there are no guarantees. Be sure to thoroughly check out all items that you may be receiving in your return.
- **Don't Blame the Other Party for a Bad Trade.** You can always decline a trade, so the responsibility is always yours. If you make a bad decision, learn from it, and move on.

How to Begin the Bartering Process

Although bartering is a fairly easy and straightforward process, there are some simple principles you can implement to maximize your return:

- Be proactive in identifying what you need.
- Identify suitable trading partners and/or networks.
- Make contact with a person to begin your trade. Be very clear and detailed regarding what you have and what you are looking for in return.
- Negotiate the details of the trade including where you will meet, and what you are trading.
- Don't let emotions cloud good judgment. Take a 'time out' to think about it and/or get input from a trusted source.

Final Word about Bartering

With a good plan, following the fundamental principles described above, and little effort, you can use bartering to obtain the goods and services you want or need without impeding your cash flow. The possibilities are endless!

PART V: TILAPIA BUSINESS SUCCESS

CHAPTER 18

Marketing and Selling Your Fish

Industry Overview

Innovative approaches to marketing are usually the key to financial success for smaller scale fish producers. In essence, the fish must be sold for more than it cost to grow. Regardless of the size or type of venture, marketing is an essential component and requires a plan. The information contained within this chapter will help the smaller scale fish producer to formulate a marketing plan.

Most producers would like to sell to one or two high-volume buyers, such as a processing plant or distributor. This is a good marketing strategy if you are producing large quantities of fish. However, small-scale producers are not on the same economic level as larger producers are and, therefore, must usually sell for a higher price to remain profitable. Their best option is to establish niche markets for their products.

Niche markets have advantages and disadvantages. The main advantage is that producers become wholesalers and, in some cases, retailers. Consequently, producers have more control over the prices they set for their products, and they retain some portion of the profit that otherwise would have gone to middlemen. The main disadvantage is that considerable time must be spent analyzing and developing these markets. A number of critical factors should be analyzed before marketing begins.

* Competition
* Product Forms
* Price
* Type of Promotion
* Unique Ideas for Market Share
* Where to Market the Product
* Regulations

Dominating the Competition

The seafood industry has been well established and can be competitive in some areas. Small and medium size aquaculture (as well as aquaponic) operators compete with wild-caught and large farm-raised fish of both domestic and foreign origin. This isn't necessarily a bad thing. A matter of fact, armed with the right knowledge and presentation, it is a positive that you can use to your advantage.

Understanding your competition helps you develop production and marketing programs and markets that will provide the highest profits. Also, remember that other seafood products are not the only competition you will have to contemplate. You must consider competition from all protein products such as poultry, beef, and pork.

However, the market for fish is vast and the demand for mercury-free, healthy fish is increasing at an exponential pace. This particular market, if targeted properly, can present an environmentally minded, healthy fish farming entrepreneur with endless opportunities. Furthermore, the price of fish and meat is increasing at an astonishing rate.

The U.S. consumption of fish continues rise. As more consumers become concerned with the preservation of the environment, initiatives are launched to improve the way we use resources. This is integral to food production in America. Monterey Bay Aquarium's Seafood Watch Program announced that Chinese-raised Tilapia as an "avoid" in a watch list for seafood. Interestingly, tilapia imports from China account for 62 percent of US tilapia imports. Tilapia harvests in China, by far the world's largest producer, were estimated by the United Nations to be 1.8 million tons in 2019. The U.S. buys about a third of China's tilapia.

Only about 5% of tilapia is produced domestically in the United States. The remaining 95% of tilapia consumed in the is imported. However, as consumers are becoming more aware of the dangers of consuming foreign and ocean caught fish, the demand for organically raised American fish becomes even greater. Since domestic fish farmers do not have the capacity to meet the demand, excellent opportunities exist for domestic operators to gain market share and develop a profitable industry.

Use all of your resources—industry experts, the telephone book, and your own energy—to help evaluate the competition. Talk with potential customers to determine their level of interest. Realize that the development of a new market requires substantial effort.

Product Forms

A unique product form can make your business stand out. The size of the product can also be important to the selected market. One of the best ways to select a product form is to find out what the customers want and give it to them. For instance, channel catfish are usually sold after reaching a live weight of one to two pounds. At this size, a 1.5-pound fish will yield two 4.5-ounce fillets.

The following is a list of the more common fish product forms, with descriptions and specific information relatedto each one.

- **Live** fish are sold to live-haulers who stock fee-fishing lakes or farm ponds, or sell to consumers who dress them
at home for consumption.
- **Fish in the round** are put on ice and sold just as they come out of the water.
- **Drawn** fish have their entrails removed and are usually sold on ice.
- **Dressed** fish are sold completely cleaned with the entrails removed. Heads may be left intact, as trout are often sold,but generally the head is removed. Fins and tails may be removed or left intact. Species such as channel catfish have the skin removed. On trout and other scaled fish, the skin is usually left intact.
- **Steaks** are cross sections of dressed fish around 1-inch thick. Larger catfish (more than three pounds) are sometimes sold as steaks.
- **Nuggets** come from the belly flap after it is cut free from the fillet. Channel catfish nuggets are common in supermarkets. Their popularity may be a result of the lower price. In general, these nuggets have a stronger flavor than fillets.
- **Fillets** are boneless pieces of fish.
- **Flank** fillets are the two sides of the fish cut away from the backbone. Rib bones and skin are usually removed.
- **Butterfly** fillets are the two, skin-on flank fillets held together by the belly flap or across the back

(with the backbone removed). Trout are sometimes sold as butterfly fillets.
- **Strips** are smaller pieces of fish cut from fillets. Strips are usually breaded, marinated, or used for other value-added treatments.
- **Deboned** fish have the rib and back bones removed, with the rest of the body intact.
- **Smoked** fish is a value-added product. Two smoking methods (hot and cold smoking) are employed. Hot smoking never produces enough drying to ensure safe keeping without refrigeration.
- **Hot Smoking** involves temperatures of 250° to 300°F for a period of four to five hours. Cold smoking, on the other hand, preserves fish by drying. Cold smoking requires as little as 24 hours or as long as three weeks at temperatures never exceeding 80°F. If you decide to get involved in smoking, there are a number of potential regulations that must be addressed. Proposed FDA regulations are described in subpart A of 21 CFR part 123 of the Federal Register.
- **Breading** fish also adds value (and weight) to a fish product. The fish are generally dipped in liquid batter (usually milk or egg mixtures) and rolled in seasoned breadcrumbs or corn meal.

The most common processed product forms are dressed, fillets, nuggets, and steaks. The preferred product size will depend on an individual customer's preferences. Fillets, for example, are generally cut into prescribed proportions that yield a single serving (four to eight ounces) from one or two fillets. As a rule, the whole fish needs to be at least 1.25 to 2.5 pounds to obtain the appropriate size fillets. The dress-out percentage, or yield, on fish such as channel catfish, hybrid striped bass, tilapia, or trout range from 33 percent on some fillets to more than 60 percent for whole dressed fish. Frozen or refrigerated are also forms that need to be considered.

TABLE 27. **Tilapia Dressed Out (Fillets)**

TILAPIA SPECIES	FISH WEIGHT AT HARVEST (G)	YIELD (% OF TOTAL WEIGHT)	MEAT YIELD (G)
Oreochromis niloticus	143.5 ± 18.2	34	84.3
Puntius gonionotus	545.3 ± 32.2	52.2	341.3
Oreochromis niloticus	165.6 ± 21.2	39.2	105.3
Puntius gonionotus	503.7 ± 30.1	48.2	351.1
Oreochromis niloticus	124.7 ± 15.1	29.5	77.3
Puntius gonionotus	219.8 ± 25.1	21	154.1
Oreochromis niloticus	433.8 ± 16.2	34.2	266.9
Puntius gonionotus	1268.8 ± 30.5	40.5	846.5

Regardless of the product form you choose to offer, it is especially important to establish and maintain a reputation for quality and reliability. Be sure to gain an accurate understanding of each customer's needs before delivering the first fish.

Pricing for Success

Putting a price on your product is not as simple as you might think. Often, pricing a product is an agonizing, lengthy decision that will likely require periodic adjustments to reflect internal and external circumstances, as well as new market environments.

The lowest price to charge would be equal to your cost per pound, including both fixed and variable costs. The highest price would be what a customer could be talked into paying. The following are a number of factors to consider when establishing a product's price:

- How will the product be positioned in the food fish market? Is it more like caviar or carp (i.e., local sustainably raised premium organic fish vs. fish caught off Japan/China in heavily polluted waters laced with mercury, Fukushima radiation, and other contaminants?
- Who are the customers? What are they accustomed to paying? Are they individual consumers, up-scale restaurants, or food wholesalers?
- What species and prices are competitors offering?
- What quality perceptions and uniqueness, if any, are associated with the chosen species or culture method?

Many systems have been developed to aid in pricing products. The following are descriptions of the systems most relevant to small-scale marketing.

- **Cost-Plus** pricing simply adds a constant percentage of profit above the cost of producing a product. The problem
- with cost-plus pricing is that it is difficult to accurately assess fixed and variable costs. This pricing system works fine in the absence of severe competition.
- **Competitive pricing** is probably the easiest and, in retail marketing, the most common form of pricing. In this system, producers gather market information on prices and quantities of competing products and then price their products accordingly.
- **Skimming** involves introducing a product at a relatively high price for more affluent, quality-conscious customers. Then, as the market becomes saturated, the price is gradually lowered.
- **Discount pricing** offers customers a reduction from advertised prices for specific reasons. For example, a fish farm advertises in the local newspaper that prices will be 25 percent less if they bring the advertisement from the paper. Or a producer who advertises on local radio offers customers a discounted price when they mention the advertisement. Discount pricing can often apply to purchases of larger quantities.
- **Loss-Leader** pricing is offering a limited selection of the products, at a reduced price, for a limited time. The goal is to attract more customers to the producer's place of business so that they might also buy non-discounted products as well. This pricing method is seen at farmers markets and supermarkets to introduce a new product or to create consumer interest.
- **Psychological** pricing involves establishing prices that look better or convey a certain message to the buyer. For example, instead of charging $7.00 per pound, the producer charges $6.99 per pound. This will make the product appear to be more of a bargain. Or, instead of charging a price close to production costs, the producer charges a higher price that buyers associate with a higher quality or a more desirable fish species.
- **Perceived-Value** pricing is positioning and promoting a product on non-price factors such as quality, organic, farmed sustainably, or grown locally. Then, the producer must decide on a price that reflects this perceived value. An example of this strategy would be promoting local, organic

raised versus imported fish or any species you could portray as having a high probability of being contaminated with mercury.

Effective Promotion and Advertising Methods

After a product and price have been decided upon, a promotional strategy needs to be developed. Promotion is a way to attract customers. Ideally, a high-quality product in demand will sell itself. However, if no one is aware that the product is for sale (when and where), no sales will be made. Time allocated to promoting a product typically results in a worthwhile payoff. The two general methods of promoting fish products are generic and personal promotions.

- **Generic** promotion is commonly performed by large commodity groups such as The Catfish Institute, the National Aquaculture Association, etc. This type of advertising promotes a certain type of product but does not endorse any particular brand or company.
- **Personal** promotion is used to distinguish your product from other products. A number of methods of personal promotion are available to small-scale fish marketers. Word-of-mouth advertising is one of the best types of personal promotion. One customer who is satisfied will tell their friends about your product. The multiplying effect of word-of-mouth promotion can be tremendous, but often slow and further promotion will be required. It is also important to remember that a dissatisfied customer will also tell friends.

Other common channels for advertising include radio, newspaper, TV, magazines, handbills, flyers, and posters. The promotional message must be clear, to the point, and focused.

Point-of-purchase materials, such as recipes and information about your aquaponic operation, will help

maximize sales. Before creating your own point-of-purchase materials, decide if available materials can be adapted for your use. For example, a variety of recipes and general information are available about farm-raised fish through the library, internet, public agencies, and various associations.

Social media is the interaction among people in which they create, share, or exchange information and ideas in virtual communities and networks. Social media marketing can help spread the word about your business, quality products, and services. Internet users continue to spend more time with social media sites than any other type of site.

Social media market size in 2021 in the United States is a project to be a $61.4 billion dollar industry. Furthermore, social media is expected to have a 16.5% growth rate over 2020.

Smartphones and other mobile devices have become more widespread, 28% of American adults now report that they go online "almost constantly. Overall, 85% of Americans say they go online on a daily basis (year 2021). That figure includes the 35% who go online almost constantly, as well as 48% who say they go online several times a day and 8% who go online about once a day.

Social media marketing is useful for every business that has customers who are using the internet. It is wise to bring in outside help to establish a social media marketing strategy that can be carried out and implemented. Many firms and qualified individuals will do your social media marketing for you at a wide range of prices and levels, reliant upon your needs and budget. Effective social media marketing places your product in front of your target audience.

The form(s) of promotion you choose will depend on the scale of your operation, available resources, availability of the product, and geographic location of the operation. In addition to public advertising, it is important to consider on-site product promotion, both visual and verbal. Remember to include the non-price attributes of the product that will help develop repeat customers. The following is an itemized list of potential marketing strategies.

- Point-of-Purchase Materials
- Flyers
- Posters
- Word-of-Mouth
- Craigslist
- Clubs and Organizations
- Auto Wrap
- Business Alliances
- Farmer's Markets
- College Campuses
- Facebook and Twitter Account
- Company Website
- LinkedIn
- Social Media
- Newspaper and local shopping guide advertisements
- Local, State, Federal Association Memberships

Unique Ideas for a Large Market Share

The small aquaculture operator often finds it necessary to provide some unique product or service to carve out a piece of the market. This uniqueness can be obtained by providing a custom order service, offering special delivery schedules, providing products not readily available in the area, etc. Be careful not to commit to any schedules or make promises that can't be fulfilled. Where the product is marketed can provide some interesting possibilities. The best approach is emphasizing that you have locally grown healthy tilapia, free of toxins.

Best Places to Market Your Product

There are many different marketing and product outlets for the small-scale aquaponic operator. Your choices will be affected by costs, such as processing, delivery, advertising, overhead, materials, equipment, and personal time. Species selection, product form, target market, and company location will also have a profound effect on this issue. Selling the fish to a large processor is often not desirable or possible for the small operator. This does not mean, however, that there are not available markets.

Direct Retail Sales

Direct retail sales, where the producer sells directly to the customer, is generally where the greatest per-unit profit is realized. Direct retail sales to consumers is a good place to start if supplies are small or availability of the product is uncertain. The following is a list and description of several direct retail sales options.

- **Local Customer Base:** This is the simplest of all direct marketing options. Individual sales are made to customers on a repeat basis. Clients pickup from you or you deliver. A customer base takes time to develop, but using advertising materials such as the local newspaper, Craigslist, or a direct mailer containing news on availability,

new products, nutrition information, and recipes can speed-up the process.
- **Roadside Market:** This option has many variations. The product can be live, fresh iced, or in some cases, dressed and iced. A small market may be operated at the farm site, or a live tank can be set up at a more populated location with heavier traffic. The fish may be kept in a live tank on the truck or in a tank set up at the remote location, with the permission of the property owners. Off-farm locations may include busy intersections, convenience stores, gas stations, farmers markets, flea markets, or liquor stores. The mobile marketing technique brings the product to the people and increases the potential market area. Check with local officials to determine if permits or other restrictions apply.
- **Fish Fry Fund-Raiser:** Many groups use a fish fry to raise money. They include churches, schools, hospitals, civic groups
- (Scouts, YMCA, Women's Clubs, Trail Life USA, Christian Brigade, Lion/Rotary/Elks/etc.,), political groups, and other non-profit organizations. Marketing to these groups may require larger quantities of fish of similar size. You may provide just the fish products or cater the entire event for a percentage of the ticket sales. Other aquaculture product catering opportunities include events such as birthday parties, weddings, and other private parties.
- **Office Building Markets:** Tall buildings hold lots of people who go home from work hungry, but often don't want to stop at the store or fish market. Contacts are made in offices through bulletin boards, flyers, word-of-mouth, and direct sales. A sales force can even be recruited from clerical workers who can make sales during the course of the workday. Sales can even be made during the early part of the week with deliveries later in the week. Ice, coolers, and individual packing are required for this type of marketing.
- **Fairs and Festivals:** This is a proven marketing option. County and state fairs are excellent target markets. A list of these events can generally be obtained from the local or state chamber of commerce. These events draw hungry crowds. Much of the food is overpriced and not particularly good. Good, healthy fish plates provide an opportunity to capitalize on this opportunity, as well as promote and educate the public on the benefits of aquaponics. On the downside, often a commission or fee is paid to the fair organization.
- **Value-Added Market:** Each of these marketing techniques could be considered value-added if the fish are processed to customer specifications. Other value-added products include smoked, breaded, or marinated fish. Customers in this market understand that they will need to pay premium prices for quality products and services.
- **Tank Harvest Sale:** This is a popular marketing technique that works very well for both small and medium size operations. By planning ahead and advertising in local papers and radio, a farmer may be able to sell an entire crop in one day. Prepare holding facilities for sale of any left-over fish.
- **Bartering:** Trading fish for other products and/or services is a great way for all parties to get what they need with-out incurring a tax burden. Carpet cleaning, landscape work, painting, beauty care, automotive service, trading for fruits, etc. are viable alternative markets for obtaining true value from your product.
- **Direct Wholesale Sales:** Wholesaling to other businesses that sell directly to the consumer is another option. Although, the direct-wholesale option usually reduces the per-unit profit it can increase the units sold.

Set up appointments with managers of every restaurant, grocery store, and food wholesaler within a 50-mile radius of the production site. Find out beforehand, if possible, individual preferences for species, product form, size, volume, availability, and prices. Have a strong sales pitch prepared and a fresh sample of your products. Pricing in the wholesale market is usually based on individual negotiation, so determine your price range and have a negotiation strategy.

Some managers will be immediately interested while others will not. For those who are interested, customize the product to fit individual needs. Keep your customers satisfied by sup- plying the size, form, quantity, and quality of product that the customer expects. Especially important are good human relations skills, showing common courtesy, and building rapport. A list and description of several direct wholesale options follow.

- **Live Hauling:** Live haulers generally purchase fish at the site and transport them to other outlets, including processing plants, pay lakes, recreational lakes, or retail outlets. Small-scale producers often have difficulty working with live haulers because the producers lack the large quantities of fish the haulers need to make it a profitable activity. There are, however, companies that charge a fee for custom harvesting. These companies are generally in large pro- duction areas, and it may be difficult to get them to a small pro- duction facility. Live haulers prefer not to handle small quantities of fish (less than 1,000 to 2,000 pounds, and in some areas not less than 5,000 to 10,000 pounds). One advantage of selling to a live hauler is that there is no additional personal investment of time or equipment to process, transport, or sell your fish.
- **Sales to Local Restaurants:** Restaurants can be an excellent market for fish farmers. Growing fish to match the desired plate portion as well as the weekly volume can be a good source of revenue. The typical restaurant will take ten to 80 pounds of fish per week. Restaurants like unique and new items for their "catch of the day" menu. Learning to produce a popular fish species and marketing it out of season can bring big dividends. Work with a chef to develop a new dish using your product. It is good advertising for both you and the restaurant.

  When deciding which businesses to contact, remember that many businesses serving food are not necessarily identified as "restaurants." Do not overlook the country club, the VFW, caterers, or the corner pub.

  Once a restaurant becomes a customer, make a point of helping to educate the staff about your product. Educating the head of the serving staff and providing a short brochure, or other printed information, may be a key to continued success.
- **Supermarkets:** Many seafood markets and supermarkets buy locally produced fish. Retail chain supermarkets offer a good market for larger quantities of fish. Unless a supermarket is locally owned and operated, it might be necessary to supply part or all of the chain stores. This may be more volume than the small operator can handle. A number of the large superstore markets now have live fish tanks that need a consistent supply of quality live fish (20 to 50 pounds per week). Smaller supermarkets and seafood stores are generally easier to work with, and more likely to sell local products. Educating the staff about your products in these settings is also extremely important. It is a good idea to offer point-of-sale information for use at the sea- food counter.
- **Specialty Stores:** These stores include ethnic grocery stores, gourmet shops, and health food stores. Fish is an important part of people's diets in some cultures. Health food stores may be willing to try your product because the perceived quality and healthy aspects of farm-raised products is usually higher than that of wild-caught fish. Ethnic markets are usually more willing to purchase whole fish. Each of these markets has special demands for equipment, capital, time, and effort.

Licenses, Permits, and Regulations for Raising and Selling Fish

Fish production and marketing activities are regulated at the local, state, and even federal levels. Depending on your operation, health inspections as well as business and sales tax permits may be required. Following are health permit sources for several kinds of operations.

- Retail outlets and restaurants: County Health Department
- Processing facilities: Governed by most state public health agencies and regulated by local jurisdictions (i.e., county health departments, etc.).
- In addition to a state's Fish and Game agency, many states also have Food and Drug Branch that oversee fish farming.
- Interstate commerce: U.S. Food and Drug Administration and the USDA (United States Dept. of Agriculture).

Checking with the chamber of commerce (local regulations), city hall (business licenses), and your state's Department of Conservation and Natural Resources ("privilege" license for operating a fish wholesaling establishment) may be a good idea during the formulation of your business and marketing plan. For retail sales, a sales tax permit may be required. For mobile operations, the Department of Transportation should be consulted. You can also check with your county's Extension Office or Fisheries Specialist regarding this inspection policy.

Unfortunately, "Big Brother" can get so involved with oversight, requirements, policies, and periodic inspections along all points of the operation that running a business can be challenging to the small operator, due to the additional time drain caused by these regulatory requirements. Operators must keep records of their testing, meet various operating requirements, and record every detail about their business. With all this regulatory oversight and associated cost in accommodating so many different agency requirements, many operators choose to run their operations under the radar.

Marketing Plan Synopsis

Before any production begins, it is prudent to establish a solid marketing plan. One of the primary considerations in developing this plan is the time and effort that you can devote to marketing your product. Evaluate the market for the best species options, keeping in mind your personal situation (including finances, experience, and time availability). In addition to planning, critical issues such product, price, promotion, and location must be evaluated prior to the onset of production. One of the best ways to address these and other critical questions discussed is to develop a written marketing plan.

This plan should include many of the same items as a business plan. The plan should detail goals, financial data including capital required, budgets, and cash flow analysis, how regulatory requirements will be met, a detailed list of necessary equipment, and a feedback system to monitor the progress of the venture. Haphazard business planning will lead to an inefficient and possibly even a failed enterprise. A detailed plan provides direction and helps avoid some of the pitfalls associated with any new venture.

Regardless of the market avenues chosen, it is best to target specific markets. Determine what size market you can service well and limit your initial marketing program to those areas. Develop more than one market outlet. The key to niche marketing success is to develop and maintain a reputation for quality, dependability, and excellent customer service. Rewarding business opportunities are always open to those creative individuals who are willing to plan, work hard, and persist.

PART V: TILAPIA BUSINESS SUCCESS

Appendix

APPENDIX

Helpful Resources

FarmYourSpace.com *Website Topics*

TABLE 28.

#	"FARMYOURSPACE" WEBSITE CATEGORIES	#	"FARMYOURSPACE" WEBSITE CATEGORIES
1	6 Easy Bread Recipes Healthy & Homemade	25	Human Urine as Fertilizer
2	7 Reasons to Begin Aquaponics	26	Hydroponics
3	8 Reasons Why to Grow Your Own Food	27	Importance of Supporting Decentralized, Small-Scale Farming
4	8 Ways How to Make Healthy Organic Soil	28	Make your own bell siphon
5	10 loans or grants you can use to obtain property	29	Maximizing Your Space Tips
6	Animals / Poultry / Livestock	30	Money Making Ideas
7	Aquaponic Business Plans	31	Money Saving Tips
8	Aquaponic Design Plans	32	Mothballs to Keep Snakes Away
9	Backyard Aquaculture	33	Non-toxic Cleaning
10	Backyard chicken coops for different climates and budgets	34	Nutrition, Health, Organics
11	Beekeeping	35	Off-grid Living
12	Canning	36	Raised Bed Gardening Tricks
13	Clever Gardening Ideas	37	Root Cellars
14	Composting	38	SUPERFOODS You Can Grow
15	Cooking	39	Survival
16	Couple earns six figure Income	40	Surviving Big Brother & Gov't Regulations
17	Do-It-Yourself Drawings & Instructions	41	Sustainable Farming
18	Fruit & Nut Trees	42	The Many Health Benefits of Gardening
19	Gallery of Gardens (submit page)	43	Time to Reclaim our Food Independence is Now
20	Grow 25 pounds of Sweet Potatoes in a Bucket	44	Tips for starting your own Organic Garden
21	Homesteading 101	45	Traditional Gardening (New Ideas!!!)
22	How A Home Garden Produce Enough Food To Live On	46	Vertical Gardening
23	How to make your own DIY fertilizer	47	Water Conservation & Water Quality
24	How to Quickly Pickle Veggies	48	Worms (Vermiculture)

Farm Your Space

The author is creating a website **www.FarmYourSpace.com** that will have a great deal of helpful information on it, a Q&A feature where you can get your questions answered, many articles, pics and videos related to aquaponics, vertical gardening, and other helpful ideas to maximize your space; as well as improving the effectiveness and efficiency of your farming operation. I encourage you to check it out. Table 43 lists some of the topics FarmYourSpace.com will cover:

NOTE: Some popular aquaponic resources commonly found on the Internet are not included within the following list of resources, as the author has found that they either do not provide sound advice, lack integrity, or are not reliable enough to recommend.

The Aquaculture Network Information Center

The Aquaculture Network Information Center (AquaNIC) is gateway to the world's electronic resources in aquaculture AquaNIC is maintained at Purdue University in West Lafayette Indiana, and is supported by the Illinois-Indiana Sea Grant Program and Purdue University's Department of Animal Sciences. The AquaNIC site contains links to aquaculture sit at other state Land Grant universities, USDA sites devoted aquaculture, professional organizations, and other sites with aquaculture information. **AquaNIC can be accessed via: WWW: http://ag.ansc.purdue.edu/aquanic/**

CropKing, Inc.

CropKing, Inc., has been specializing in the business of controlled environment agriculture and hydroponics since 1982 and manufactures greenhouse structures at their facility in Lodi, OH. The company sells to both hobby and commercial growers throughout the United States as well as internationally, emphasizing quality, competitive pricing, and a full range of services to its customers. For more information visit: www.cropking.com; email: cropking@cropking.com; or call 330-302-4203; 134 West Drive, Lodi, OH 44254

Professional Aquaculture Services

559 Cimarron Drive
Chico, CA 95926
PAC is operated by Tony Vaught, with over 30 years experience in production aquaculture. PAC offers a *source of fish* in their Aquaponics System Starter

Package for those establishing a new aquaponics system, consulting services and trouble-shooting consultation for system design, feeding and fish health.
Phone: (530) 343-0405
Cell: (530)-519-1051
Fax: (530) 343-0405
tvaught@proaqua.com
http://www.proaqua.com/
http://www.aquaculturedirect.com/

Useful Internet Web Links for More Information

- www.iasproducts.com/Main.html
  (Has good prices for some aquaculture components, including auto feeders)
- www.caaquaculture.org/
  (The California Aquaculture Association site)
- www.dfg.ca.gov/Aquaculture/
  (California Dept. of Fish & Game site for aquaculture, permit included on disk)
- www.fish.washington.edu/wrac/
  Western Regional Aquaculture Center website
- http://aqua.ucdavis.edu/index.htm
  (The UC Davis Aquaculture website has a lot of publications and links)

Greenhouses

- *How to Build Your Own Greenhouse*, Roger Marshall, ISBN: 13-978-1-58017-587-6

SARE Learning Center

The USDA's Sustainable Agriculture Research and Education (SARE) program has funded the publication of many fine books over the years and now offers some of them as free downloadable PDFs. For example, *Building a Sustainable Business* is $17 in print but free as a download. SARE also publishes bulletins, grant project reports, and much other useful information. www.sare.org/publications/.

Note to Reader

Please visit the website **"FarmYourSpace.com"** to obtain other helpful information and post your project to help others in the community.

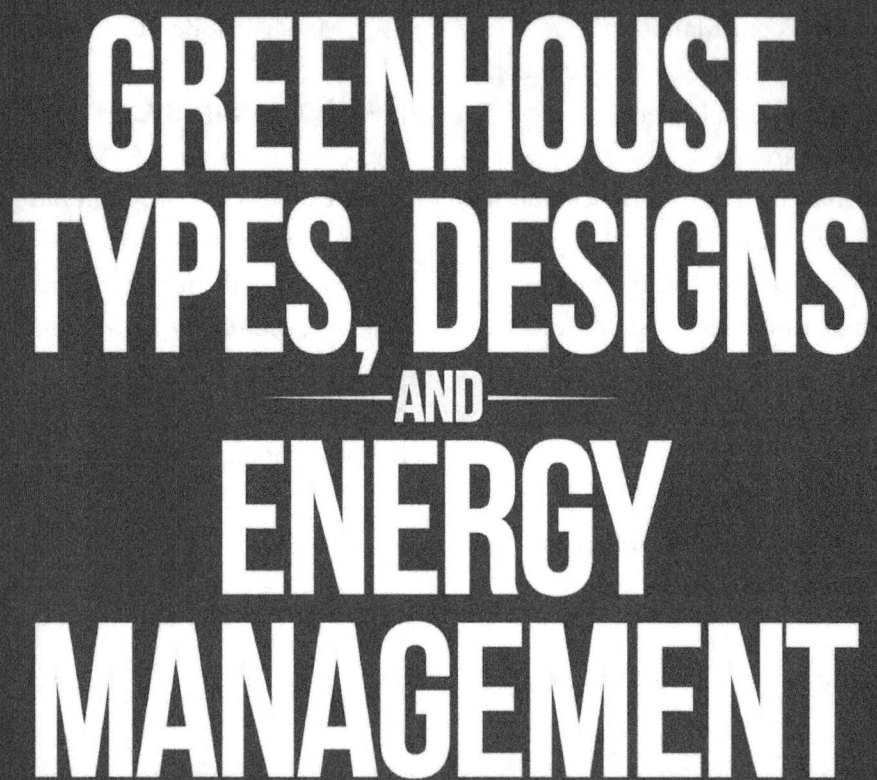

EVERYTHING YOU NEED TO KNOW ABOUT GREENHOUSES

- ✓ What you need to know before building a greenhouse.
- ✓ Things to consider before you buy a greenhouse.
- ✓ How to earn extra money with your greenhouse
- ✓ Tips for owning a hobby greenhouse.
- ✓ Easy plants for greenhouse starters.
- ✓ Watering considerations for your greenhouse.
- ✓ How to minimize energy cost.
- ✓ Things to include in your greenhouse (ergonomic tables, specialty lighting, designated spaces, etc.).
- ✓ Greenhouse shading.
- ✓ Fans, heaters, ventilation systems.
- ✓ Automation systems for your greenhouse.
- ✓ Different types of greenhouses and greenhouse materials.

Greenhouse types, construction, heating, and cooling; shading, seeds & seedlings, environmental control systems, permitting, greenhouse uses, reducing operational costs, and much more is covered in this valuable book, providing you with everything you need to know about greenhouses and achieving optimal success with your greenhouse.

AQUAPONICS
Build and Operation Manual

STEP-BY-STEP INSTRUCTIONS
400+ PAGES, 200+ HELPFUL IMAGES

| MEDIA/ GROW BED SYSTEMS | NFT SYSTEMS | RAFT SYSTEMS | HYBRID SYSTEMS |

David Dudley, PMP, PE

GROW YOUR OWN HEALTHY LOW COST FOOD, BARTER/SELL SURPLUS

- EASY TO FOLLOW BUILD DRAWINGS
- USER-FRIENDLY INSTRUCTIONS
- FUN, REWARDING, AND INTERESTING TO ALL
- FRESH ORGANIC PRODUCE AND PLENTIFUL HEALTHY FISH
- ABUNDANCE OF FISH AND VEGETABLES THROUGHOUT THE YEAR
- LOWER FOOD COST FINANCIAL REWARDS

This 400+ page user-friendly book (how-to-guide) includes:

✓ Everything You Need to Know About Aquaponics

✓ How to Set-Up & Operate an Aquaponic System of Any Size

The author, **David Dudley, PMP, PE** is a professional aquaponics consultant who has helped many individuals and institutions develop aquaponics systems. His accomplished career in aquaponics, hydroponics, and aquaculture includes serving as the Construction Manager of the Oklahoma Aquarium, Engineering Manager of the nation's largest caviar producing company, overseeing life support systems of four large aquaculture facilities, designing a $5M aquaculture operation for white sturgeon, and Project Manager of a large fishing clinic facility for the U.S. Department of Wildlife.

David also holds advanced degrees in civil engineering and nutrition/dietetics, operated a small commercial nursery for 10+ years, and has several decades of experience in vegetable gardening. David understands every facet of aquaponics and clearly communicates aquaponics in a way that truly helps others.

www.FarmYourSpace.com

ISBN 978-0-9998304-7-5

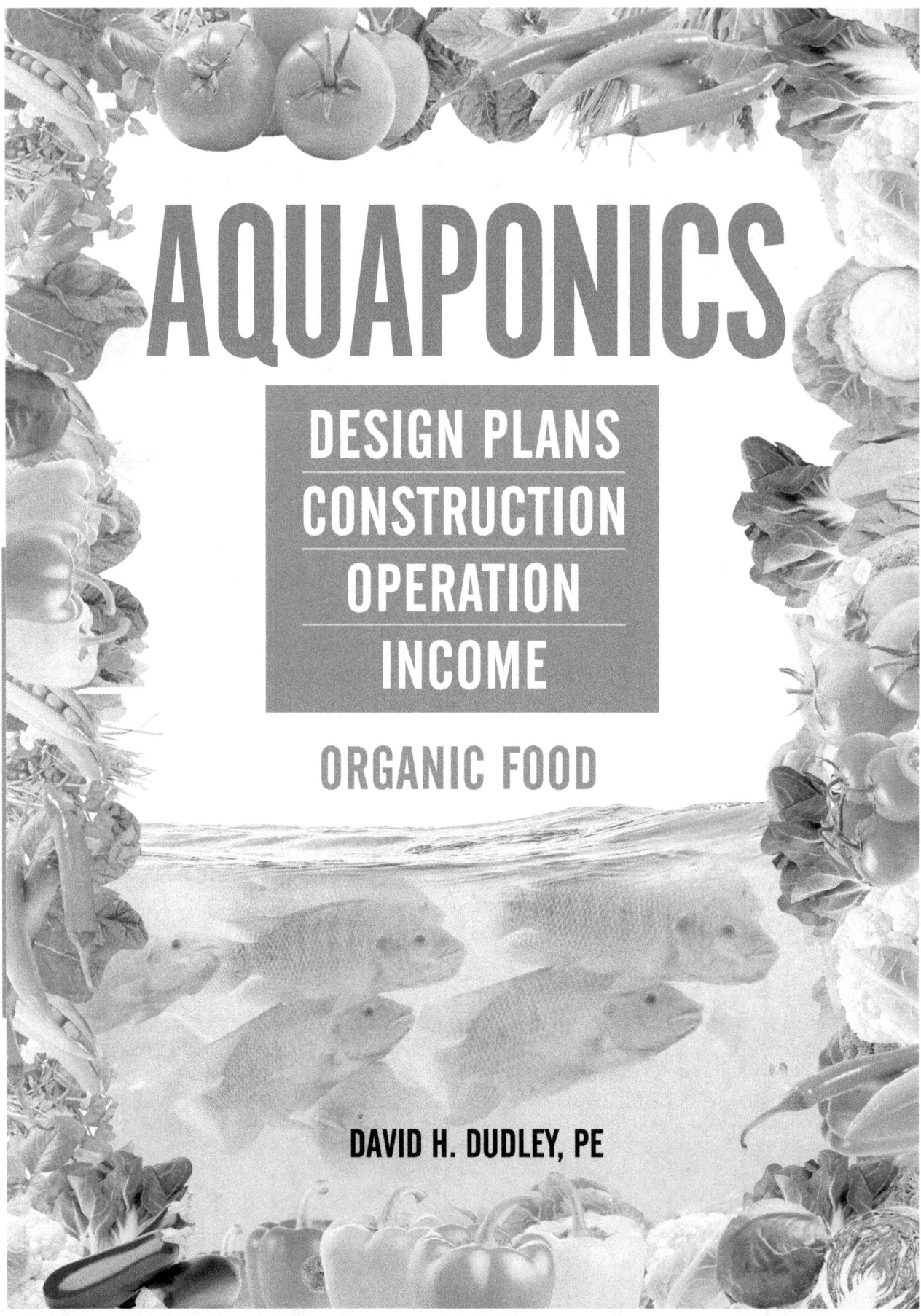

AQUAPONICS

DESIGN PLANS
CONSTRUCTION
OPERATION
INCOME

ORGANIC FOOD

GROW YOUR OWN HEALTHY FOOD
SAVE MONEY
EARN EXTRA INCOME

BEGINNER TO COMMERCIAL
SMALL TO LARGE
AQUAPONIC SYSTEMS

This 625+ page easy to follow comprehensive book provides illustrations, step-by-step instructions, and real life photos showing you how to do everything.

This 625+ page user-friendly book (how-to-guide) provides:

✓ EVERYTHING YOU NEED TO KNOW ABOUT AQUAPONICS
✓ HOW TO SET-UP & OPERATE AN AQUAPONIC SYSTEM OF ANY SIZE
✓ HOW TO OBTAIN FINANCIAL REWARDS FROM YOUR AQUAPONICS SYSTEM

Within these sections you will be provided everything you need to know, in an understandable, easy to follow approach; so that you can enjoy environmentally-friendly sustainable farming, consistently feed your family plentiful healthy organic lost cost food, and earn as much extra income as desired. Best of all, this book will show you how to accomplish these objectives in the most efficient way possible. Aquaponics truly is a worthwhile and rewarding endeavor.

The author, **David H. Dudley, P.E.,** is a professional aquaponics consultant who has helped many individuals and companies develop aquaponics systems. His accomplished career in aquaponics, hydroponics, and aquaculture includes serving as the Construction Manager of the Oklahoma Aquarium, Engineering Manager of the nation's largest caviar producing company, overseeing life support systems of four large aquaculture facilities, designing a $5M aquaculture operation for white sturgeon, and Project Manager of a large fishing clinic facility for the U.S. Department of Wildlife.

David also holds advanced degrees in civil engineering and nutrition/dietetics, operated a small commercial nursery for 10+ years, and has several decades of experience in vegetable gardening. David understands every facet of aquaponics and clearly communicates aquaponics in a way that truly helps others.

www.FarmYourSpace.com

AQUAPONICS FOR PROFIT

EARN **EXTRA MONEY** OR CREATE A **SUCCESSFUL COMMERCIAL BUSINESS**

What are the best methods for making money with aquaponics? Which edible aquaponic fish species will generate the most revenue per pound? Which aquaponic vegetables provide the highest profit margin? Which fish and vegetables are in highest demand by consumers? Do you know about all of the non-vegetable plants that can provide you with greater revenue than vegetable plants? What costs are involved in setting-up and operating an aquaponic system? What is the cost-benefit analysis of an aquaponic system? Where and how can I sell my aquaponic harvest? What regulations and legalities are involved in setting-up and operating an aquaponic business? Did you know that there are other aquatic species that can be grown in aquaponics which can generate more revenue than the commonly grown edible aquaponic fish species? What is the best way to barter my aquaponic harvest? How can I get my products officially labeled as 'organic'? What is the best approach to having a successful aquaponics business that will produce the largest profit margin? How can I earn extra money with my small backyard aquaponic system? Which type of aquaponic system – Media-Bed/Flood-and-Drain, Nutrient Film Technique (NFT), Raft System / Deep Water Culture (DWC) – is the most profitable? What are the pros and cons to each of these types of aquaponic systems? In addition to selling your harvest, are you aware of all the other ways in which you can earn revenue from your aquaponic system? How much time would I need to invest to have an aquaponic system that will feed my family and provide us with some extra income?

This user-friendly easy read book will answer the above a questions. This valuable resources is also packed with the necessary information that will not only show you how to make extra money with your aquaponic system, but to grow it into a successful commercial business; if that is your desire. Also included are two real-world aquaponic business plans. This book is an excellent investment that will reward you greatly with the knowledge needed to earn extra money through aquaponics or optimize revenue from a commercial aquaponic operation.

David H. Dudley is a professional aquaponics consultant who has helped many individuals and companies develop aquaponics systems. His accomplished career in aquaponics, hydroponics, and aquaculture includes serving as the Construction Manager of the Oklahoma Aquarium, Engineering Manager of the nation's largest caviar producing company, overseeing life support systems of four large aquaculture facilities, designing a $5M aquaculture operation for white sturgeon, and Project Manager of a large fishing clinic facility for the U.S. Department of Wildlife. David also holds advanced degrees in civil engineering and nutrition/dietetics, owns a commercial nursery, and has several decades of experience in vegetable gardening. David understands every facet of aquaponics and clearly communicates aquaponics in a way that truly helps others.

www.FarmYourSpace.com

APPENDIX

AQUAPONICS
HOW TO DO EVERYTHING
2nd Edition

FROM BACKYARD TO PROFITABLE BUSINESS

David H. Dudley, PE

AQUAPONICS
HOW TO DO EVERYTHING
FROM BACKYARD TO PROFITABLE BUSINESS

FRESH ORGANIC VEGETABLES AND PLENTIFUL HEALTHY FISH

FEED YOUR FAMILY HEALTHY LOW COST FOOD, BARTER and/or SELL SURPLUS

Beginning Basics to Operating a Profitable Commercial Business

Expensive university courses and lengthy on-site training workshops that cost thousands of dollars do not provide as much valuable, comprehensive material, as presented in this a user-friendly 'how-to' book.

This **550+ page book** (how-to-guide) consists of three very important sections:

- ☑ Everything You Need to Know About Aquaponics
- ☑ How to Set-Up & Operate an Aquaponic System of Any Size
- ☑ How to Successfully Operate a Profitable Commercial Aquaponics Business

Within these sections you will be provided everything you need to know, in an understandable, easy to follow approach; so that you can enjoy environmentally-friendly sustainable farming, consistently feed your family plentiful healthy organic lost cost food, and earn as much extra income as desired (the sky really is the limit). Best of all, this book will show you how to accomplish these objectives in the most efficient way possible. Aquaponics truly is a worthwhile and rewarding endeavor.

The author, **David H. Dudley, P.E.**, is a professional aquaponics consultant who has helped many individuals and companies develop aquaponics systems. His accomplished career in aquaponics, hydroponics, and aquaculture includes serving as the Construction Manager of the Oklahoma Aquarium, Engineering Manager of the nation's largest caviar producing company, overseeing life support systems of four large aquaculture facilities, designing a $5M aquaculture operation for white sturgeon, and Project Manager of a large fishing clinic facility for the U.S. Department of Wildlife.

David also holds advanced degrees in civil engineering and nutrition/dietetics, owns a commercial nursery, and has several decades of experience in vegetable gardening. David understands every facet of aquaponics and clearly communicates aquaponics in a way that truly helps others.

ISBN 978-0-9969090-7-5

APPENDIX

AQUAPONICS
PLANS AND INSTRUCTIONS
MEDIA-BED (FLOOD-AND-DRAIN) SYSTEMS

David H. Dudley, P.E.

AQUAPONICS PLANS AND INSTRUCTIONS
MEDIA-BED (FLOOD-AND-DRAIN) SYSTEMS

This 400+ page user-friendly book shows you how to easily produce an abundance of Fresh Organic Produce and Plentiful Healthy Fish through a Media-Bed (Flood-and-Drain) Aquaponic System; so you can:

Feed Your Family Healthy Organic Food, Barter and/or Sell Surplus, Substantially Lower Your Food Cost.

This VALUABLE resource has everything from Beginner Basics to showing you how scale-up as you desire to grow your system. Easy to follow step-by-step Instructions and SO much more.

Expensive university courses and lengthy on-site training workshops which cost thousands of dollars do not provide as much valuable, comprehensive material as presented in this comprehensive user-friendly 'how-to' book.

Included are Media-Bed Design Plans, Instructions & Everything You Need to Know about Aquaponics.

Along with the instructions are over 350 photos and illustrations which show you how to set-up and operate a productive and successful media-bed aquaponic system of any size; and how to scale-up in size to produce even more organic vegetables and fish as you desire grow.

This book empowers you with the knowledge needed to consistently feed your family environmentally friendly sustainable healthy organic food and greatly lower your food bill. Fun, Financially Rewarding, Enjoyable, Healthy, and an Interesting Conversation Topic at social functions.

David H. Dudley is a professional aquaponics consultant who has helped many individuals and companies develop aquaponics systems. His accomplished career in aquaponics, hydroponics, and aquaculture includes serving as the Construction Manager of the Oklahoma Aquarium, Engineering Manager of the nation's largest caviar producing company, overseeing life support systems of four large aquaculture facilities, designing a $5M aquaculture operation for white sturgeon, and Project Manager of a large fishing clinic facility for the U.S. Department of Wildlife. David holds advanced degrees in civil engineering and nutrition/dietetics, owns a commercial nursery, and has several decades of experience in vegetable gardening. David understands every facet of aquaponics and clearly communicates aquaponics in a way that truly helps others.

www.FarmYourSpace.com

- Aquaponic Plans and Instuctions
- Fresh Organic Produce and Plentiful Healthy Fish
- An Abundance of Fish and Vegetables throughout the Year at a Very Low Cost
- User-Friendly Plans and Instructions from Beginner to Scaling-Up to Any Size
- Aquaponics is Fun and a Conversation Topic Others Find Very Interesting.
- An Excellent Investment with Tremendous Personal and Financial Rewards.

APPENDIX

Aquaponic Design Plans
Everything You Need to Know
from
Backyard to Profitable Business
2nd EDITION

David H. Dudley, PE

Aquaponic Design Plans, Instructions & All You Need to Know
Fresh Organic Produce and Plentiful Healthy Fish
Feed Your Family Healthy Food + Barter and/or Sell Surplus
Everything from Beginner Basics to Operating a Profitable Aquaponic Business

Expensive university courses and lengthy on-site training workshops which cost thousands of dollars do not provide as much valuable, comprehensive material as presented in this comprehensive user-friendly 'how-to' book.

Aquaponic Design Plans
Everything You Need to Know *from*
Backyard to Profitable Business
2nd EDITION

This 546-page book provides detailed directions to create and maintain different types of aquaponic systems of all sizes so you can consistently feed your family environmentally friendly sustainable healthy organic food and earn extra income. This valuable how-to resource consists of three important sections:

* Design Plans, Instructions & Everything You Need to Know About Aquaponics
* How to Set up & Operate different types of Aquaponic Systems of any Size
* How to Turn Aquaponics Into a Profitable Venture

The author, David Dudley, is a professional aquaponics consultant who has helped many individuals and companies develop aquaponics systems. His accomplished career in aquaponics, hydroponics, and aquaculture includes serving as the Construction Manager of the Oklahoma Aquarium, Engineering Manager of the nation's largest caviar producing company, overseeing life support systems of four large aquaculture facilities, designing a $5M aquaculture operation for white sturgeon, and Project Manager of a large fishing clinic facility for the U.S. Department of Wildlife. David holds advanced degrees in civil engineering and nutrition/dietetics, owns a commercial nursery, and has several decades of experience in vegetable gardening. David understands every facet of aquaponics and clearly communicates aquaponics in a way that truly helps others.

www.FarmYourSpace.com

ISBN: 978-0-9985377-4-0

APPENDIX

APPENDIX

AQUAPONICS
FLOOD-AND-DRAIN MEDIA BED SYSTEMS

SMALL TO LARGE AQUAPONIC SYSTEMS

GROW YOUR OWN HEALTHY LOW COST FOOD, BARTER/SELL SURPLUS

This 400+ page user-friendly book (how-to-guide) includes:

- ✓ Everything You Need to Know About Aquaponics
- ✓ How to Set-Up & Operate an Aquaponic System of Any Size
- ✓ How to Obtain Financial Rewards from Your Aquaponics System

Everything you need to know, described in an easy to follow approach; so that you can immediately start feeding your family lots of healthy organic food, lower your food bill substantially, earn extra income, and enjoy the environmentally-friendly benefits of sustainable farming. This book will show you how to accomplish these objectives in the most efficient way possible. Everyone will be eager to learn more about your aquaponics gardening operation. Aquaponics truly is a fun, worthwhile, and rewarding endeavor. You can also barter or give away your harvest surplus.

The author, **David H. Dudley, P.E.,** is a professional aquaponics consultant who has helped many individuals and institutions develop aquaponics systems. His accomplished career in aquaponics, hydroponics, and aquaculture includes serving as the Construction Manager of the Oklahoma Aquarium, Engineering Manager of the nation's largest caviar producing company, overseeing life support systems of four large aquaculture facilities, designing a $5M aquaculture operation for white sturgeon, and Project Manager of a large fishing clinic facility for the U.S. Department of Wildlife.

David also holds advanced degrees in civil engineering and nutrition/dietetics, operated a small commercial nursery for 10+ years, and has several decades of experience in vegetable gardening. David understands every facet of aquaponics and clearly communicates aquaponics in a way that truly helps others.

www.FarmYourSpace.com

ISBN 978-0-9998304-7-5

BUSINESS AND ENTREPRENEUR SUCCESS MANUAL

DAVID H DUDLEY

BUSINESS AND ENTREPRENEUR SUCCESS MANUAL

Everyone desires to be successful. With all the various success tips we are regularly bombarded with you'd think the world would be populated with successful people. Why then do some achieve it and others don't? What are the defining factors that separate those that are highly successful from those that only achieve mediocre success?

There are certain rules, principles, characteristics, and methods that are have proven to be effective for all those who have achieved success. When you have the right skills and pit them into practice, you will be successful in both your personal and professional life. This book is designed to fill your tool box with the tools essential for success. In order to optimize success in your life, you simply have to move forward implementing the valuable user-friendly practical information provided within the book.

The author, **David H. Dudley,** has created and grown several successful businesses in his career, and most recently consolidating them into one successful enterprise. Along the way, he has enjoyed success in several different career fields, beginning with nutrition/dietetics, and from there transitioning into engineering and aquaponics. His decades of research and study into what it takes to be successful have equipped him with the ability to accomplish his goals efficiently.

He now shares that information in this user-friendly valuable manual to help you achieve the level of success you desire in all areas of your life.

APPENDIX

COUNTRY BREEZE FARM

Overview

Country Breeze Farm is a holistic community with a heart that empowers persons with intellectual disabilities to live to their fullest potential with dignity and purpose. We provide exceptional housing, life-enriching programs, and loving, comprehensive care in a safe, vibrant, inclusive farm-style community.

Our residents and day clients, called Farmers, are valued as individuals. Each is important, personally known by caregivers and staff, and nurtured to live a happy and full life. Farmers live, work and socialize with their peers on the farm, and also have lots of opportunities to interact with the local community, and participate in a wide variety of activities.

Country Breeze Farm is anchored in the core values of faith, hope, love, compassion, and all care is delivered by the principles of the Golden Rule. These values become part of each Farmer, caregiver and employee and are manifest in a friendly community that is filled with remarkable and authentic joy and growth. New milestones, miracles, and expanded capabilities of the Farmers are witnessed every day; and are often celebrated in one way or another so as to provide positive reinforcement and further embolden a community team atmosphere. It's a place where residents and day clients become family with each other and staff, and a place where Country Breeze Farm becomes home.

Mission

Country Breeze Farm is a vibrant not-for-profit community for adults with cognitive disabilities, and often, other challenges. We are dedicated to enabling our residents, and day clients, to achieve their full potential, providing them and their families a future of hope. Country Breeze Farm encompasses a safe and comfortable rural environment with programs and services that enables our special farmers to be blessed with a good quality life with purpose, love, and nurturing companionship.

www.CountryBreezeFarm.org

Country Breeze Farm

Secure Campus & Facilities

Country Breeze Farm has a wide variety of facilities and infrastructure to support our special ladies and gentlemen, called Farmers. First, consider that the farm has a completely secure campus featuring a security station at the entrance and a secure perimeter fence. The security station has full gate control and provides video monitoring for anyone entering or leaving the premises.

24-Hour Video Monitoring by Staff and Loved Ones

In addition to stringent protocol and well-trained staff, video monitoring is provided at various places around the farm, which are accessible to management at all times. Some of the live video feeds can be viewed by families online.

Residential Housing with Community Areas

The living arrangements feature individualized bedrooms custom furnished for each resident according to their special needs. The open community room has tables, couches, and recliners where residents can hang out, relax, share meals or snacks, socialize, watch a show on the screen or play games. The facility is staffed 24/7, features specialized bathrooms, and safety glass throughout.

Community Building

Country Breeze Farm features a unique multi-use building with a stage that allows for worship services, cafeteria dining, indoor recreational activities, and entertainment events such as plays and musicals. The building also has a fitness center, classrooms for arts and education, administrative offices, a kitchen, a reception area, bathrooms, and maintenance and storage facilities.

Recreation & Events

Country Breeze Farm features a campus designed and filled with peaceful places, outdoor lunch & picnic areas, and more. The outdoor recreation areas include places for bocce ball, basketball, volleyball, frisbee, soccer, walking paths, fishing, paddleboats & canoeing, picnic outings, as well as places for music concerts and festivals. Indoor recreational events and activities are just as plentiful and include a fitness facility, music, as well as many other activities.

Indoor Riding Arena & Barn for Animals

Country Breeze Farms features the opportunity for Therapeutic Horseback Riding. This allows residents and day clients the opportunity to enjoy and bond with the animals. An outdoor fenced-in area is also available for riding horses and keeping goats.

Greenhouses & Aquaponics

Greenhouses and aquaponics on-site allow our special ladies and gentlemen to actively work at producing healthy food, raise fish, and grow flowers.

Outdoor Organic Farming

Healthy food production extends out to the fully functional outdoor organic gardening areas. Residents, day clients, and staff thoroughly enjoy eating the produce they grow, and the satisfaction of selling it at the Country Breeze Market.

Chicken House & Homing Pigeons

Chickens and homing pigeons provide further animal interaction. Fresh eggs and are available for our special ladies and gentlemen to eat and sell.

Environmentally Friendly Campus

Country Breeze Farms is an environmentally friendly campus. It features permeable driving and parking areas with natural filtration and buffers to mitigate stormwater runoff and pollution. Buildings are also energy efficient and contain appliances and equipment that reduce the impact on the environment. The facility emphasizes recycling, conservation, and the use of environmentally friendly products.

Programs

Country Breeze Farm encompasses a safe and comfortable environment with programs and services that enables our residents and day clients (called farmers) to be blessed with a good quality life with purpose, love, and nurturing companionship. Our around the clock qualified loving staff help our farmers discover and enjoy occupational projects, whether it be caring for the land and animals, cultivating and harvesting organic produce and flowers, helping in the office, or working in other ways on campus. Well-trained staff encourage and assist everyone so they can participate in many different activities, such as therapeutic horseback riding, exercising in the fitness facility, crafting, music program, indoor/outdoor recreational activities, excursions into town, speech/occupational therapy, and daily chores.

APPENDIX

Country Breeze Farm

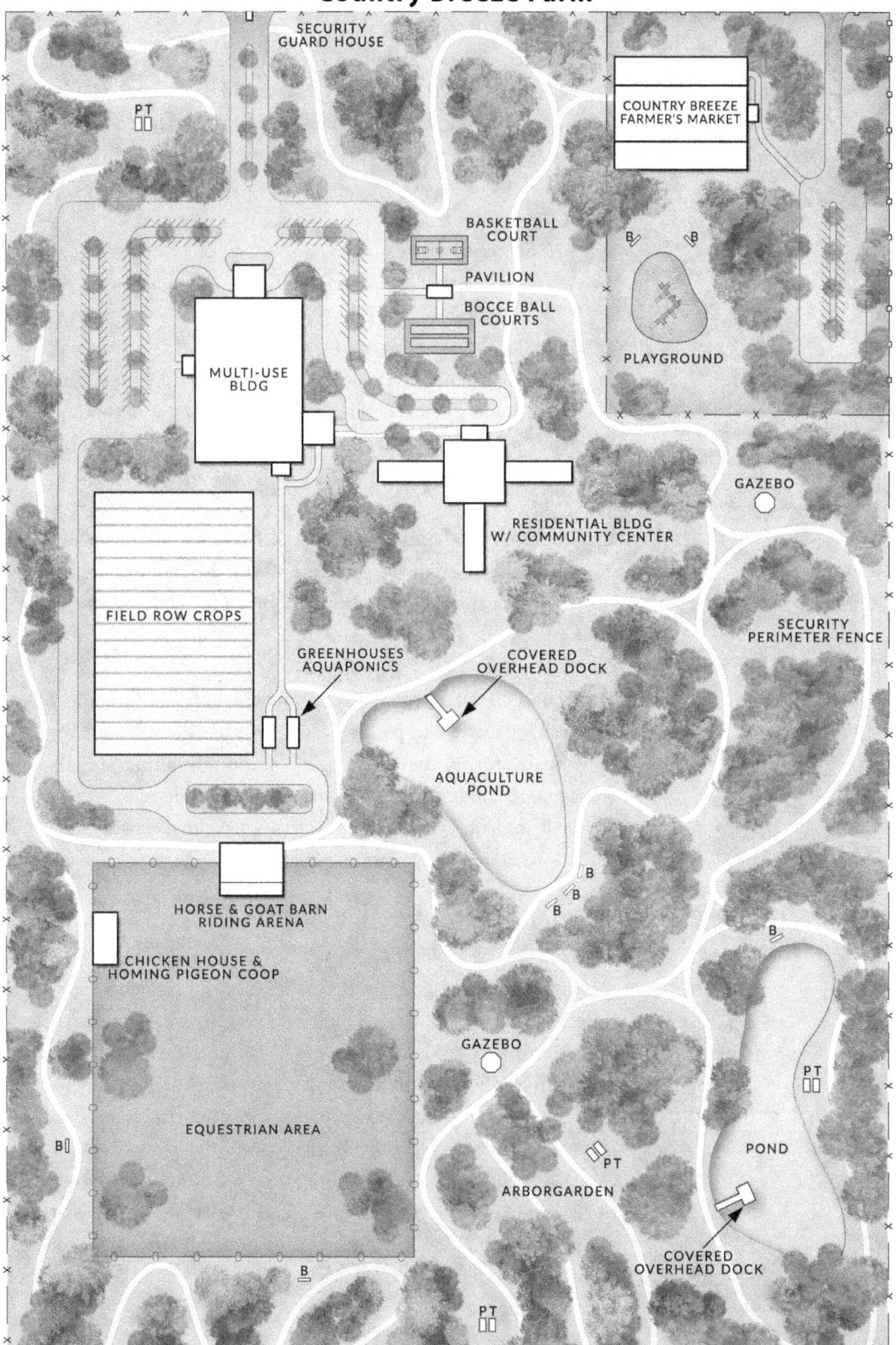

If you benefit from this book, would you please leave me a positive review on Amazon? It only takes a moment. Positive reviews are a tremendous help to me and are greatly appreciated.

Thank you SO much!

www.ingramcontent.com/pod-product-compliance
Lightning Source LLC
Chambersburg PA
CBHW081153070526
44583CB00021B/2815